Coefficient of Variation and Machine Learning Applications

Coefficient of Variation and Machine Learning Applications

K. Hima Bindu

Raghava M

Nilanjan Dey

C. Raghavendra Rao

CRC Press
Taylor & Francis Group
Boca Raton London New York

CRC Press is an imprint of the
Taylor & Francis Group, an **informa** business

CRC Press
Taylor & Francis Group
52 Vanderbilt Avenue,
New York, NY 10017

First issued in paperback 2021

© 2020 by Taylor & Francis Group, LLC
CRC Press is an imprint of Taylor & Francis Group, an Informa business

No claim to original U.S. Government works

ISBN-13: 978-0-367-27328-6 (hbk)
ISBN-13: 978-1-03-208419-0 (pbk)

**Visit the Taylor & Francis Web site at
http://www.taylorandfrancis.com**

**and the CRC Press Web site at
http://www.crcpress.com**

Contents

List of Figures, ix

List of Tables, xi

Preface, xiii

Authors, xvii

CHAPTER 1 ▪ Introduction to Coefficient of Variation   1

  1.1   INTRODUCTION   1

  1.2   COEFFICIENT OF VARIATION   3

  1.3   NORMALIZATION   5

      1.3.1   Coefficient of Variation of Normalized
             Variable   5

      1.3.2   Illustration   7

      1.3.3   Random Variable with Probability Density
             Function   8

             1.3.3.1   *Normal Distribution with Mean*
                        $\mu(\neq 0)$ and Standard Deviation σ   8

             1.3.3.2   *Gamma Distribution with Mean*
                        μ and Standard Deviation σ   8

      1.3.4   Random Variable with Probability Mass
             Function   8

1.4 PROPERTIES OF COEFFICIENT OF
     VARIATION   9

     1.4.1 Properties of Mean   9

     1.4.2 Properties of Standard Deviation   9

     1.4.3 Properties of CV   9

     1.4.4 Features Based on Coefficient of Variation   10

            1.4.4.1 *Influence of Translation and Scale
                     on Features of CV*   11

     1.4.5 CV of Mixture Distributions   13

            1.4.5.1 *Mixture of Normal Distributions*   16

1.5 LIMITATIONS OF COEFFICIENT OF
     VARIATION   22

1.6 CV INTERPRETATION   22

1.7 SUMMARY   23

1.8 EXERCISES   24

CHAPTER 2 ▪ CV Computational Strategies   27

2.1 INTRODUCTION   27

2.2 CV COMPUTATION OF POOLED DATA   28

     2.2.1 Illustration   30

2.3 COMPARISON OF CV WITH ENTROPY
     AND GINI INDEX   31

2.4 CV FOR CATEGORICAL VARIABLES   32

     2.4.1 Table Lookup Method   33

     2.4.2 Mapping Method   34

     2.4.3 Zero Avoiding Calibration   35

2.5 CV COMPUTATION BY MAPREDUCE
     STRATEGIES   36

2.6 SUMMARY   38

2.7 EXERCISES   38

CHAPTER 3 ▪ Image Representation 43

3.1 INTRODUCTION 43

3.2 CVIMAGE 44

    3.2.1 CVImage Representation of Lena Image 45

3.3 CV FEATURE VECTOR 48

    3.3.1 Demonstration 50

    3.3.2 Hvalue Distribution Analysis 51

    3.3.3 Ranking of CV Features 54

3.4 SUMMARY 55

3.5 EXERCISES 55

CHAPTER 4 ▪ Supervised Learning 57

4.1 INTRODUCTION 57

4.2 PREPROCESSING (DECISION ATTRIBUTE CALIBRATION) 58

4.3 CONDITIONAL CV 59

4.4 CVGAIN (CV FOR ATTRIBUTE SELECTION) 61

    4.4.1 Example 61

4.5 ATTRIBUTE ORDERING WITH CVGAIN 61

4.6 CVDT FOR CLASSIFICATION 63

    4.6.1 CVDT Algorithm 64

    4.6.2 CVDT Example 64

    4.6.3 Using CVDT for Classification 67

4.7 CVDT FOR REGRESSION 68

4.8 CVDT FOR BIG DATA 71

    4.8.1 Distributed CVDT Induction with Horizontal Fragmentation 71

    4.8.2 Distributed CVDT Induction with Vertical Fragmentation 72

4.9   FUZZY CVDT   73

4.10 SUMMARY   76

4.11 EXERCISES   76

CHAPTER 5 ▪ Applications   79

5.1   IMAGE CLUSTERING   79

5.2   IMAGE SEGMENTATION   82

5.3   FEATURE SELECTION   82

5.4   MOOD ANALYSIS   84

        5.4.1   Bipolar Disorder   84

        5.4.2   Twitter Mood Predicts the Stock Market   84

5.5   CV FOR OPTIMIZATION   85

5.6   HEALTH CARE   86

5.7   SOCIAL NETWORK   87

5.8   SUMMARY   87

5.9   EXERCISES   87

APPENDIX A, 89

REFERENCES, 121

INDEX, 125

List of Figures

Figure 1.1   Influence of *translation* and *scale* on *mean*   12

Figure 1.2   Influence of *translation* and *scale* on
standard deviation   12

Figure 1.3   Influence of *translation* and *scale* on *CV*   13

Figure 1.4   Influence of *translation* and *scale* on
normalized mean   14

Figure 1.5   Influence of *translation* and *scale* on
normalized standard deviation   14

Figure 1.6   Influence of *translation* and *scale* on
normalized CV   15

Figure 1.7   Mixture distributions example (f_1 has
$\mu = 4, \sigma = 1$; f_2 has $\mu = 6, \sigma = 2$)   16

Figure 1.8 α Ratio 1:1   17

Figure 1.9 α Ratio 1:2   18

Figure 1.10 α Ratio 2:1   18

Figure 2.1   Binomial distributed variable   32

Figure 2.2   Normal distributed variable   33

Figure 2.3   Binomial distribution with zero avoidance   35

Figure 3.1   CVImage illustration: (a) Pixels values,
(b) CVImage demo, and (c) CVIF demo   46

Figure 3.2   CVImage: (a) Sample image, (b) CVImage
of sample image, and (c) CVImage with
$CVT = 33\%$ for edge representation   47

Figure 3.3 CVImage of Lena: (a) Lena image and its variants; (b) CVImage with CVT = 3, 9, 15, 21; (c) CVImage with CVT = 27, 33, 39, 45; and (d) CVImage with CVT = 51, 57, 63, 69   48

Figure 3.4 Color image and histograms of its color channels   49

Figure 3.5 SampleDB   51

Figure 4.1 CVDT for data in Table 4.2   67

Figure 4.2 Regression tree of AutoMPG data set   68

Figure 4.3 Variation of RMSE with alpha   70

Figure 4.4 RMSE of CVDT and CART   71

Figure 4.5 CVDT for fuzzy membership of *Play = Yes*   74

Figure 4.6 CVDT for fuzzy membership of *Play = No*   75

Figure 4.7 Fuzzy membership influence on RMSE and ESS   75

Figure 4.8 Two-dimensional data for classification   78

Figure 5.1 Elbow method on color images to determine the optimal number of clusters with CV Features. (a) Elbow method on 9 CV features. (b) Elbow method on 3 CV features   80

Figure 5.2 Purity of clusters. (a) Purity with 9 CV Features. (b) Purity with 3 CV features   81

Figure 5.3 Image segmentation with CVImage and CVThreshold. (a) Brain tumor image; (b) result of segmentation with CVThreshold = 3; and (c) result of segmentation with CVThreshold = 9   83

List of Tables

Table 1.1	Mixture of 2 normals with mixing α ratio 1:1	19
Table 1.2	Mixture of 2 normals with mixing α ratio 1:2	20
Table 1.3	Mixture of 2 normals with mixing α ratio 2:1	21
Table 2.1	Data for exercise 1	38
Table 2.2	Empirical distribution of data for exercise 1	38
Table 2.3	Super market transactions for exercise 3	40
Table 3.1	Image statistics	50
Table 3.2	Hash values	52
Table 3.3	Hash value distribution of sample DB	53
Table 3.4	Image databases	53
Table 3.5	Hvalue distribution database wise	54
Table 4.1	Train data for play tennis	58
Table 4.2	Play tennis data	60
Table 4.3	Decision table for Outlook = Sunny	66
Table 4.4	Decision table for Outlook = Overcast	66
Table 4.5	Decision table for Outlook = Rain	66
Table 4.6	Play tennis data with fuzzy memberships	74
Table 4.7	Decision system	76
Table 4.8	Employee data set	77
Table 4.9	Regression data set	78
Table A.1	CV features of Greenlake	91

Table A.2   CV features of Greenland in gray scale
(Hvalue = 1)   94

Table A.3   CV features of TrainDB   98

Table A.4   CV features of TestDB   109

Table A.5   CV features of Indonesia   114

Table A.6   CV features of SampleDB   116

Table A.7   CV features of Testimages   117

Table A.8   CV features of Images_Texture   119

Preface

THIS BOOK *Coefficient of Variation and Machine Learning Applications* is an outcome of the experiences of the authors. It is observed that several statistical measures like mean, standard deviation, and moments are adopted as features for understanding and representing systems. The coefficient of variation (CV), the percentage relative variation to its mean, has received less attention in the domain of Data Mining and Machine Learning.

Coefficient of variation has been used for making decisions in domains like Economics, Business Administration, and Social Science. Coefficient of Variation indicates the constancy aspects of the system (data). It is a potential candidate for consideration as one of the features for representing a system/object/subject. Coefficient of variation can also be used as an index to rank order or compare the systems/objects/subjects in place of Gini Index (measure of inequality), Entropy (measure of disorder), etc. Coefficient of variation will support in designing computationally efficient single pass algorithms due to its elegant algebraic properties.

The book is organized into five chapters. It is intended to provide introduction, computational strategies, representation, some of the Machine Learning and Data Mining methodologies, and few potential applications.

Chapter 1 is an introduction to Coefficient of Variation. The reader can learn basic ideas, definitions and computation of CV for a data vector and a random variable. It describes normalization and transformation issues. Properties of CV, features that can be derived from CV, possible interpretations based on CV and its limitations are discussed in this chapter. This chapter sets stage for subsequent chapters.

Chapter 2 presents the distributed computational strategies of CV using pooled and Map-Reduce approaches. It also presents the pre-processing method to be used for categorical data to enable CV computations. This helps CV to be applicable for real-world data used in Machine Learning.

Chapter 3 proposes using CV as an operator for producing CVImage as a derived representation for an image. It presents image binarization by tuning a threshold on CV. Using features derived based on CV of the RGB colors, a CVFeature vector as a knowledge representation mechanism is developed, which has interesting applications.

Chapter 4 introduces supervised learning using CV. It defines CVGain as an attribute selection measure of classification and regression trees. The suitability of CVGain for distributed environments and fuzzy decision systems is discussed.

Chapter 5 presents applications like image clustering and segmentation by using CVImage features. It discusses CV-based feature ordering and feature selection, leading to a candidate order in all Data Mining and Machine Learning approaches. Details of how CV is potential to perform Mood analysis, provide solution for optimization problem are outlined in this chapter. It ends with few pointers to address health care and social network applications.

This book can be used as a concise handbook as well as a part of courses in Data Mining, Machine Learning and Deep Learning. The authors feel that this book can help researchers in solving Machine Learning applications with its decent coverage of illustrations and demonstrations.

Raghavendra Rao C. expresses his special thanks to Mr. K. Omprakash, General Manager—Education Sales, Capricot Technologies Private Limited for providing Maltab academic license which is used in generating some of the results and figures.

Hima Bindu would like to acknowledge the Director—Prof. C. S. P. Rao, the In-charge Registrar—Prof. G. Amba Prasad Rao and faculty members of Computer Science and Engineering department at NIT Andhra Pradesh for their encouragement and cooperation.

The authors thank Mr. M. Ramakanth, BSNL, for valuable inputs related to interval estimations of CV. The efforts of Mr. Adithya Chillarige and Dr. Tilottama Goswami are acknowledged for sharing the images and image databases.

Authors

Dr. K. Hima Bindu is an alumnus of NIT—Warangal, JNTU—Hyderabad, and University of Hyderabad. She has a decade of experience in Data Mining and Machine Learning. She is currently serving as an assistant professor in National Institute of Technology, Tadepalligudem, India. Her research interests include Data Mining, Machine Learning, Big Data, and Educational Data Mining.

Dr. Raghava M obtained his MTech from Mysore University, Karnataka, India, during 2003. He received his PhD from University of Hyderabad, Telangana, India. He started his engineering teaching career in CVR College of Engineering, Hyderabad, India, during 2003 and successfully handled various courses related to Systems Engineering, Neural Networks, Data Engineering, and Linux Internals. Currently he is serving as a professor in CVR College of Engineering. His areas of interests include Computer Vision, Regularization Theory, Sparse Representations, and Graphical Models.

Nilanjan Dey is an assistant professor in Department of Information Technology at Techno India College of Technology (under Techno India Group), Kolkata, India. He is a visiting fellow of University of Reading, London, UK, and visiting professor of Duy Tan University, Vietnam. He was an honorary visiting scientist at Global Biomedical Technologies Inc., California (2012–2015). He is a research scientist of Laboratory of Applied Mathematical Modeling in Human Physiology, Territorial Organization of Scientific and Engineering Unions, Bulgaria, and associate researcher of Laboratoire RIADI, University of Manouba, Tunisia. He is a scientific member of Politécnica of Porto. He was awarded his PhD from Jadavpur University, West Bengal, India, in 2015. In addition, recently he was awarded as one among the top 10 most published academics in the field of Computer Science in India (2015–2017) during "Faculty Research Awards" organized by Careers 360 at New Delhi, India. Before he joined Techno India, he was an assistant professor of JIS College of Engineering and Bengal College of Engineering and Technology, West Bengal, India.

He has authored/edited more than 50 books with Elsevier, Wiley, CRC Press, and Springer, and published more than 300 papers. His h-index is 32 with more than 4900 citations. He is the editor-in-chief of *International Journal of Ambient Computing and Intelligence* (IJACI, IGI Global, UK, Scopus), *International Journal of Rough Sets and Data Analysis* (IGI Global, US, DBLP, ACM dl), co-editor-in-chief of *International Journal of Synthetic Emotions* (IGI Global, US, DBLP, ACM dl), and *International Journal of Natural Computing Research* (IGI Global, US, DBLP, ACM dl). He is the series co-editor of Springer Tracts in Nature-Inspired Computing (STNIC), Springer; series co-editor of Advances in Ubiquitous Sensing Applications for Healthcare (AUSAH), Elsevier; series editor of Computational Intelligence in Engineering Problem Solving and Intelligent Signal processing and data analysis, CRC Press (FOCUS/Brief Series); and Advances in Geospatial Technologies (AGT) Book Series (IGI Global), US. He serves as an editorial board member of several international journals, including *International Journal of Image Mining* (IJIM), Inderscience, associated editor of IEEE Access (SCI-Indexed), and *International Journal of Information Technology*, Springer.

His main research interests include Medical Imaging, Machine learning, Computer Aided Diagnosis as well as Data Mining. He has been on program committees of over 50 international conferences, a workshop organizer of 5 workshops, and acted as a program co-chair and/or advisory chair of more than 10 international conferences. He has given more than 40 invited lectures in 10 countries, including many invited plenary/keynote talks at the international conferences such as ITITS2017 (China), TIMEC2017 (Egypt) and SOFA2018 (Romania), BioCom2018 (UK), etc.

Prof. C. Raghavendra Rao completed his BSc and MSc in Statistics from Andhra University, Andhra Pradesh, India, and Osmania University, Telangana, India, respectively, and PhD in Statistics and MTech (CS & Engineering) from Osmania University.

He started his carrier as a lecturer in Statistics at Osmania University in 1984. Since 1986, he is working in the School of Mathematics and Computer/Information Sciences, University of Hyderabad, India. Presently he is a senior professor in the School of Computer and Information Sciences, University of Hyderabad.

His current research interests are Simulation and Modeling, Rough Sets, and Knowledge Discovery.

Dr. Rao is a member of the Operation Research Society of India, Indian Mathematical Society, International Association of Engineers, Society for development of Statistics, Andhra Pradesh Society for Mathematical Sciences, Indian Society for Probability and Statistics, Society for High Energy Materials, International Rough Set Society, Indian Society for Rough Sets, and ACM. He was the founder secretory of Indian Society for Rough Sets and also a fellow of The Institution of Electronics and Telecommunication Engineers, Society for Sciences, and Andhra Pradesh Akademi of Science. He is a senior elected member for International Rough Set Society 2016.

He has over 175 journals and conference proceeding papers to his credit.

Books in his credit are:

- *Evaluation of Total Literacy Campaigns* (1995) co-authored with Prof. Bh. Krishnamurti and Dr. I. Ramabrahmam, published by Booklinks Corporation, Hyderabad.

- *Multi-disciplinary Trends in Artificial Intelligence: 8th International Workshop, MIWAI 2014*, Bangalore, India, December 8–10, 2014, Proceedings, 8875, co-editors Murty, M. Narasimha; He, Xiangjian; Weng, Paul published by Springer.

- *Strategic Polysemantic search for Web Mining* (2013) co-authored by Dr. K. Hima Bindu, Publisher Lambert Academic Publishing.

Restricted Report

"Integrated Low Fidelity Model for Fuel Air Explosive" (2013)—A Restricted Report from ACRHEM to HEMRL (DRDO), Pune submitted co-authored with Prof. Arun Agarwal, Prof. Rajeev Wankar, Dr. Apparao, and Smt. Vijay Lakshmi.

Dr. Rao contributed enormously in the project of National Interest, few to mention are:

1. Mathematical model-based Control system for swing out operation for the liquid nitrogen feeding boom for rocket launching pad in Sree Hari Kota (T.A. Hydraulics Ltd.).

2. Design of Optimal driving profile generation for diesel locomotives for Indian Railways (Consultancy project for Medha Servo Control System).

3. Mathematical modeling for 6 degree of Motion platform for building driver training simulator.

4. Mathematical modeling fuel air explosive system for Defense Research Development Organization.

5. Optimal design configuration methodologies for aerospace applications with IIT Bombay for Defence Research and Development Laboratory.

6. Modeling and Simulation of scene of offence Crime Nos. 42 & 43/2015 for Special Investigation Team.

7. Real-time data study and analysis, Medha Servo Drivers Pvt. Ltd., 2017–2018.

8. Adding intelligence to second generation ATGM, ZEN Technologies Pvt. Ltd. (2018–2019).

CHAPTER 1

Introduction to Coefficient of Variation

1.1 INTRODUCTION

The growth in communication, Internet, and computing technologies is persistently posing problems to the researchers. The data getting collected are becoming unmanageable, and developments in this direction led to the paradigm of research known as "Data Science" or "Big Data." The experience of automation for past two decades is quite encouraging and given a hope, as Machine Learning methods with computational intelligence systems are producing promising results. Hence, we are going to experience higher level of comfort, integrity, quality of life, and so on.

The sophisticated systems getting built with the help of these developments resulted in new life style, namely, e-governess, smart cities, healthcare, paperless transactions, mobility-free transactions, and so on. The smartness in performance (operations) of these systems is due to consideration of all variations in the design of the systems such that the recommended (adopted) decisions will be apt for the given context of the target state and environment.

It is time to relook at what the scientific methodologies can contribute to this futuristic systems architecture. The decision-making processes in general are associated with matching patterns that are already registered and then adopt to more or less the similar decision to current situation. The matching process is highly time consuming. To speed up this matching process, several methods are being designed and constructed. Most of these tools are based on data analysis.

Expert systems (decision support system and knowledge support system) provided tools for decision-making by encapsulating rules and built appropriate inference engines more in medical domain (Mycin [1] is an example). The tractability limitations of building knowledge and inference engines came in the way of growth in Artificial Intelligence and Expert Systems. Technological growth in databases, Data Mining, and pattern recognition tools gave a new avenue to perform knowledge discovery in databases. This led rule generation and knowledge representation in an automated manner. With an assumption that data instance is an apt representative for a particular domain, the knowledge and rules discovered can be treated as domain knowledge, which can be utilized for building inference engines. Inference engines are mainly for making decisions. Thus, a recommender system is analogous to the decision support system. As this process is around data instance, these methods are called data-centric methods.

One needs to take a decision among alternatives looking at adaptability, performance, and aptness in a given system with a fixed set of objectives, which is nothing but study of variations in performance of various strategies. A decision maker as one of the players, is trying to adopt a strategy that maximizes the profit and minimizes the loss by modeling the above as a game theory problem.

An apt ranking process of the strategies is in demand for building recommendation system. This ranking process can be addressed by various methods that are based on information theory and variational methods. The statistical measures like (information) Entropy, Gini Index, Variance, and Coefficient of Variation (CV) are in use for decision-making. Several Machine Learning contributions are mostly based on Entropy and Gini Index. The variational methods like chi square and variance are also adopted for rule discovery, testing and validation, and feature ordering and feature selection [2, 3].

Although CV is one of the popular methods used for decision-making by managers and risk analysts, it is yet to attract the attention of Data Mining and pattern recognition experts.

The variance and mean are the commonly used statistical measures to describe a system under study. Variance contributes the variation w.r.t. the location, and the mean is a location. These measures are highly sensitive to the units (mathematically, scale and translation). CV by definition is percentage of relative variation to its mean. It obeys some of the index numbers properties [4].

The Section 1.2 provides Coefficient of Variation definition and its computation. Section 1.3 discusses CV for normalized data. Sections 1.4 and 1.5 presents properties of CV and its limitations. Interpretation aspects of CV are provided in Section 1.6.

1.2 COEFFICIENT OF VARIATION

Coefficient of variation [5] abbreviated as CV is a statistical measure defined as percentage relative variation to its mean of a given data (population) given by equation 1.1

$$CV = \frac{\sigma}{\mu} * 100 \tag{1.1}$$

where σ and μ ($\mu \neq 0$) are the standard deviation and the mean of the data (population), respectively. Estimate of CV and interval estimate of CV can be found in [6].

The absolute coefficient of variation (ACV) is defined as the percentage of relative variation to its absolute mean of a given data (population) using the following equation 1.2.

$$ACV = \frac{\sigma}{|\mu|} * 100 \qquad (1.2)$$

where $| \mu |$ indicates absolute value of μ.

When X is a random variable, which can be discrete or continuous, with probability mass function $p(x)$ and probability density function $f(x)$, respectively.

Algorithm 1 returns the mean μ, standard deviation σ, coefficient of variation CV, and absolute coefficient of variation ACV for a given input data (real valued) vector X.

Algorithm 2 returns the mean μ, standard deviation σ, coefficient of variation CV, and absolute coefficient of variation ACV for a given random variable X.

Algorithm 1 Computation of Coefficient of Variation for a Given Data

---

Require: X ⊳ data vector
Ensure: μ, σ, CV, and ACV ⊳ mean, standard deviation, coefficient of variation, and absolute coefficient of variation

  1: $S \leftarrow 0$ ⊳ Sum
  2: $SS \leftarrow 0$ ⊳ Sum of Squares
  3: $n \leftarrow length(X)$
  4: **for** $i \leftarrow 1 \ldots n$ **do**
  5:      $S \leftarrow S + X(i)$
  6:      $SS \leftarrow SS + X(i) * X(i)$
  7: **end for**
  8: $\mu \leftarrow \frac{S}{n}$ ⊳ mean
  9: $v \leftarrow \frac{SS}{n} - \mu^2$ ⊳ variance
10: $\sigma \leftarrow \sqrt{v}$ ⊳ standard deviation
11: $CV \leftarrow \frac{\sigma}{\mu} * 100$ ⊳ CV
12: $ACV \leftarrow \frac{\sigma}{|\mu|} * 100$ ⊳ ACV

Algorithm 2 Computation of Coefficient of Variation for Given Probability Model (Population)

Require: Probability model of a random variable X ▷ PDF or PMF

Ensure: μ, σ, CV, and ACV ▷ mean, standard deviation, coefficient of variation, and absolute coefficient of variation

  1: $\mu \leftarrow E(X)$ ▷ expected value of X

  2: $V(X) \leftarrow E(X^2) - \mu^2$ ▷ $E(X^2)$ is expectation of X^2

  3: $CV \leftarrow \frac{\sigma}{\mu} * 100$ ▷ $\sigma = \sqrt{(V(X))}$

  4: $ACV \leftarrow \frac{\sigma}{|\mu|} * 100$

1.3 NORMALIZATION

Most of the data (Y) accessible for analytics is a transformed form of the original (X). In general, this transformation is normalized transformation (linear) such that the transformed values will be between 0 and 1. The abstract of the same is discussed below and hence the interrelationship of X and Y.

Let X be given bounded real (random) variable with a and b be the lower and upper sharp bounds. If $a \leq X \leq b$, then $r = b - a$.

r can also be computed as the least difference of arbitrary upper bound and arbitrary lower bound.

Let Y be the normalized variable of X. The Y is obtained by the below transformation:

$$Y = \frac{X - a}{b - a} = \frac{X - a}{r}.$$

Application of this transformation is known as normalization. Algorithm 3 presents the normalization procedure.

1.3.1 Coefficient of Variation of Normalized Variable

Normalization is nothing but a linear transformation

$$Y = \frac{1}{(b - a)}X + \frac{-a}{(b - a)}$$

Algorithm 3 Normalization

Require: X ▷ X is vector of real numbers or random variable
    with PDF $f(.)$ or PMF $p(.)$

Ensure: Y, a, b, r

1: $a \leftarrow \min(X)$ ▷ $\min(X)$ if X is data vector, else minimum of
    the nontrivial support of $f(.)$ or $p(.)$

2: $b \leftarrow \max(X)$ ▷ $\max(X)$ if X is data vector, else maximum of
    the nontrivial support of $f(.)$ or $p(.)$

3: $r \leftarrow b - a$

4: $Y \leftarrow \frac{X-a}{r}$

$$E(Y) = \frac{1}{(b-a)}E(X) + \frac{-a}{(b-a)}$$

($E(Y) > 0$ for non-trivial distributions)

$$V(Y) = \frac{1}{(b-a)^2}V(X)$$

Hence, coefficient of variation of Y is

$$CV(Y) = \frac{\sqrt{V(Y)}}{E(Y)} = \left(\frac{\sqrt{V(X)}}{(b-a)}\right) \Big/ \left(\frac{1}{(b-a)}E(X) + \frac{-a}{(b-a)}\right)$$

$$= \left(\sqrt{V(X)}\right) \Big/ (E(X) - a)$$

$CV(Y) > 0$ for nontrivial distributions.

$$CV(Y) = \frac{\sqrt{V(X)}}{E(X) - a} = \frac{\frac{\sqrt{V(X)}}{E(X)}}{1 - \frac{a}{E(X)}}$$

$$= \frac{CV(X)}{1 - \frac{a}{E(X)}} = \frac{ACV(X)}{D}$$

where

$$D = \begin{cases} 1 - a/E(X) & \text{when } E(X) > 0 \\ -1 + a/E(X) & \text{when } E(X) < 0 \end{cases}$$

Theorem 1: $CV(Y)$ *is a non-negative.*

Proof: $CV(Y) = \frac{CV(X)}{1-\frac{a}{E(X)}}$

Since $a \leq X \leq b$, a is always smaller than $E(X)$ (for nontrivial distributions with support $[a, b]$).

Hence, $\frac{a}{E(X)} < 1$ if $E(X) > 0$ and $\frac{a}{E(X)} > 1$ if $E(X) < 0$.

If $E(X) < 0$, then $CV(X) < 0$ and $1 - \frac{a}{E(X)} < 0$; hence ratio of these two quantities is positive. That is, $\frac{CV(X)}{1-\frac{a}{E(X)}} > 0$.

Similarly, if $E(X) > 0$, then $CV(X) > 0$ and $1-\frac{a}{E(X)} > 0$; hence, the ratio of these two quantities is positive. That is, $\frac{CV(X)}{1-\frac{a}{E(X)}} > 0$.

Hence, $CV(Y)$ is always positive. □

1.3.2 Illustration

Let X be data vector of size 25, having the values:

[0 1 -2 0 -5 3 -5 -4 1 -2 1 -3 -4 -6 0 -2 -2 0 -6 -3 1 0 -5 -5 -6]
$a = min(X) = -6$
$b = max(X) = 3$
$r = 9$
$Y = (X + 6)/9$

The normalized values are

0.67 0.78 0.44 0.67 0.11 1 0.11 0.22 0.78 0.44 0.78 0.33 0.22 0 0.67 0.44 0.44 0.67 0 0.33

$E(X) = -2.12$ (note that $E(X) < 0$)
$V(X) = 6.99$
$CV(X) = -124.67$
$ACV(X) = 124.67$
$a/E(X) = 2.83(> 1)$
$E(Y) = 0.43$
$V(Y) = 0.09$
$CV(Y) = 68.12$
$CV(Y) = \frac{CV(X)}{1-\frac{a}{E(X)}} = \frac{-124.67}{1-2.83} = \frac{-124.67}{-1.83} = 68.12$
$CV(Y) = \frac{ACV(X)}{-1+\frac{a}{E(X)}} = \frac{124.67}{-1+2.83} = \frac{124.67}{1.83} = 68.12$

1.3.3 Random Variable with Probability Density Function

Let X be random variable with non trivial support $[a, b]$, which satisfies the required coverage probability $\beta = Pr(a \leq X \leq b) = \int_a^b f(x)dx$, which is close to 1. For a choosen β, various pairs a,b are possible. Among these possible pairs, pick up a pair with minimum $b - a$ value.

1.3.3.1 Normal Distribution with Mean $\mu(\neq 0)$ and Standard Deviation σ

$CV(X) = \frac{\sigma}{\mu} * 100$

One can consider $a = \mu - k\sigma$ and $b = \mu + k\sigma$. Here the choice of k depends on the pre-defined coverage probability $\beta = Pr(-k \leq Z \leq k)$, where Z is a standard normal random variable.

$r = 2k\sigma$

$Y = (x - (\mu - k\sigma)) / (2k\sigma)$

$CV(Y) = \frac{CV(X)}{1 - \frac{\mu - k\sigma}{\mu}} = \frac{CV(X)}{1 - 1 + k\frac{\sigma}{\mu}} = \frac{100}{k}$

This CV is implicitly a function of β and independent of parental characteristics.

1.3.3.2 Gamma Distribution with Mean μ and Standard Deviation σ

One can consider $a = 0$ and $b = \mu + k\sigma$, with the choice of k depending on the coverage probability $\beta = Pr(0 \leq X \leq b)$. Then, $r = \mu + k\sigma$, $Y = X/(\mu + k\sigma)$ and $CV(Y) = CV(X)$. Which means that CV of transformed variable is same as that of the original variable. This result is obvious in normalization when $a = 0$, and CV of normalized variable is same as CV of original variable. Statistical feature CV is preserved intact in this type of normalization [5, 7].

1.3.4 Random Variable with Probability Mass Function

Let X be random variable with non trivial support $[a, b]$, which satisfies the required coverage probability $\beta = Pr(a \leq X \leq b) = \sum_{x=a}^{x=b} Pr(X = x)$ and is preferably has shorter length interval.

In case of Poisson distribution with rate parameter λ, CV $= \frac{1}{\sqrt{\lambda}}$ as the mean = variance $= \lambda$. The lower bound of the Poisson variate is zero, hence $a = 0$.

As $a = 0$ and one can consider $b = \lambda + k\sqrt{\lambda}$, the choice of k depends on the coverage probability $\beta = Pr(0 \leq X \leq b)$. Hence, $r = \lambda + k\sqrt{\lambda}$ and $Y = X/(\lambda + k\sqrt{\lambda})$, therefore $CV(Y) = CV(X)$. That is, CV of transformed variable is same as that of original variable. As $a = 0$, it is similar to that in the case of Gamma distribution. Please note that this identity will not be true for truncated (zero avoiding) or shifted distributions.

1.4 PROPERTIES OF COEFFICIENT OF VARIATION

1.4.1 Properties of Mean

Let $Y = aX + b$ be a linear transformation of X, which indicates the change of scale by a and then origin by b. Then, the mean of Y is $a \times mean(X) + b$. That is, $\mu_Y = a * \mu_X + b$, where μ_Y and μ_X are means of Y and X, respectively. Hence, the transformation applied on the feature X is to be applied on its mean to get mean of transformed variable Y.

1.4.2 Properties of Standard Deviation

Let $Y = aX + b$ be a linear transformation of X, which indicates the change of scale by a and origin by b. Then, $\sigma_Y = a * \sigma_X$, where σ_Y and σ_X are standard deviations of Y and X, respectively. Hence, standard deviation is insensitive with respective to the change of origin.

1.4.3 Properties of CV

Let $Y = aX + b$ be a linear transformation of X, which indicates the change of scale a and then origin b.

Then, the $CV_Y = \frac{a*\sigma_X}{a*\mu_X+b} \times 100 = \frac{a*CV_X}{a+\frac{b}{\mu_X}}$, where CV_Y and CV_X are coefficients of Y and X, respectively, and μ_X and σ_X are mean and standard deviation of X, respectively.

When $b = 0$, that is, when there is only scale transformation (no change of origin), then $CV_Y = CV_X$.

Coefficient of variation is positive when mean is positive and negative when mean is negative.

While absolute coefficient of variation is positive irrespective of mean being positive or negative.

Absolute coefficient of variation will be the same as coefficient of variation when the mean is positive.

Coefficient of variation is not definable when mean is zero.

In general, one can apply translation and then scaling to the data, then the transformation will be $Y_T = a(X - b)$, then

$$CV_{YT} = \frac{a * \sigma_X}{a * \mu_X - a * b} \times 100 = \frac{\sigma_X}{\mu_X - b} \times 100$$

Hence,

$$CV_{YT} = \frac{CV_X}{1 - \frac{b}{\mu_X}} \quad (1.3)$$

Equation 1.3 indicates that CV is invariant with respect to the change of scale a but variant with respect to location (translation) b. CV approaches ∞ (unbounded) as b approaches mean μ_X (assume $\mu_X > 0$) from the left. The relative translation ratio $\rho = \frac{b}{\mu_X}$ need to be bounded below by $-\infty$ and above by 1, that is, $-\infty < \rho < 1$ to have non-negative CV. In normalization, the parameter a (minimum of X) is equal to translation parameter b in the above equation. While nontrivial support $r = b - a$ is the same as scaling parameter of the above equation.

1.4.4 Features Based on Coefficient of Variation

The aggregates or statistics like min, max, range, mean, standard deviation, and moments can be calculated on features. These provide informative abstractions about a set of observations [8]. Using thresholds as cuts is an important issue in Rough Sets [9] for discretization of continuous attributes. These *cuts* provide localized information. A combination of the statistical aggregation and a set of cuts will be useful in Machine Learning and deep learning. This section considers a set of features named as "Features based on CV" (*CVFeatures*), which are used in subsequent chapters for demonstrating representation of objects as well as for learning. We will discuss image representation using CV of a sliding window in Chapter 3. Further, using threshold of 33%, we will develop *flag*, leading to a bit vector that will be used as Hash value (Hvalue) in Chapter 3. In Chapter 5, a Boolean decision system is generated to perform mood analysis.

The data can be represented by using the following features:

1. min (minimum of the data)

2. max (maximum of the data)

3. S_mean (sign of mean of the data)

4. mean (magnitude of the mean of the data)

5. std (standard deviation of the data)

6. ACV (absolute coefficient of variation)

7. NCV (coefficient of variation of normalized data)

8. flag (ACV < CV_Threshold)

9. nflag (NCV < NCV_Threshold)

The features *flag* and *nflag* depend on the choice of threshold. A required number of flags can be generated by choosing appropriate set of thresholds.

1.4.4.1 Influence of Translation and Scale on Features of CV

Figures 1.1 through 1.3 depict *mean, standard deviation*, and *coefficient of variation* and its variations/influence with respect to change of *translation* and *scale*. The data considered for these figures are uniformly distributed in the range [0, 1], *translation* values are in the range [−10, 10], *scale* is in the range [0.1, 10] with step size of 0.1. As given in Algorithm 3, normalized data have been computed for each value of translation and scale for the data used for Figures 1.1 through 1.3. One can notice the monotonic behavior of *mean* after the transformation by both of these from Figure 1.1. From Figure 1.2, it is evident that the

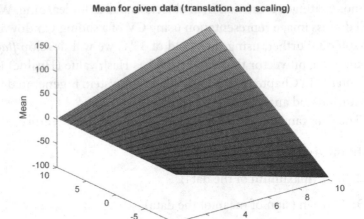

FIGURE 1.1   Influence of *translation* and *scale* on *mean*.

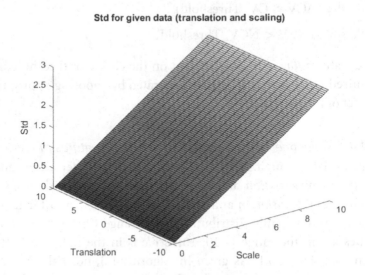

FIGURE 1.2   Influence of *translation* and *scale* on *standard deviation*.

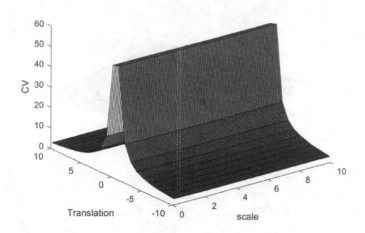

FIGURE 1.3   Influence of *translation* and *scale* on *CV*.

standard deviation is not effected due to *translation* and has linear effect with respect to *scale*. Figure 1.3 demonstrates that *scale* has no effect on *coefficient of variation*, while *translation* influences exponentially (Note that the *translation* ratio considered is $-20.2 < \rho < 20.2$). The mean, standard deviation, and coefficient of variation are computed for this normalized data and plotted from Figures 1.4 through 1.6, respectively. As all these figures resemble planes with directional cosines as $(0, 0, 1)$, z-axes as perpendicular to the plane, indicating that *translation* and *scale* have no effect for normalized data.

1.4.5 CV of Mixture Distributions

Consider k independent univariate random variables $X_1, X_2, \ldots, X_k$ with means $\mu_1, \mu_2, \ldots, \mu_k$ and standard deviations $\sigma_1, \sigma_2, \ldots, \sigma_k$, respectively. The corresponding CVs of $X_1, X_2, \ldots, X_k$ are $CV_1 = \frac{\sigma_1}{\mu_1} * 100$, $CV_2 = \frac{\sigma_2}{\mu_2} * 100, \ldots, CV_k = \frac{\sigma_k}{\mu_k} * 100$.

Let $f(x)$ be the probability density function (PDF) of Y, which is a mixture distribution of $X_1, X_2, \ldots, X_k$ with PDFs $f_1(x), f_2(x), \ldots, f_k(x)$, respectively. The respective mixing parameters are $\alpha_1, \alpha_2, \ldots, \alpha_k$ (where $\alpha_i \geq 0$ and $\sum \alpha_i = 1$).

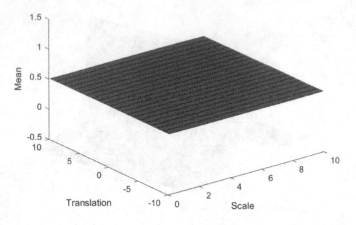

FIGURE 1.4   Influence of *translation* and *scale* on *normalized mean*.

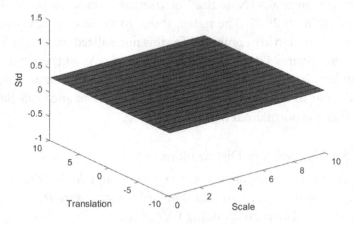

FIGURE 1.5   Influence of *translation* and *scale* on normalized standard deviation.

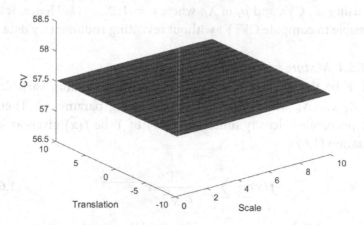

FIGURE 1.6 Influence of *translation* and *scale* on *normalized CV*.

Then, $f(x)$ is computed as given in equation 1.4

$$f(x) = \sum_{i=1}^{k} \alpha_i f_i(x) \tag{1.4}$$

CV of Y can be computed by using mean and standard deviation of X_i, $i = 1, 2, \ldots k$

Mean of Y is $\mu_Y = \alpha_1\mu_1 + \alpha_2\mu_2 + \cdots + \alpha_k\mu_k$

Expected value of Y^2 is $E(Y^2) = \alpha_1(\sigma_1^2 + \mu_1^2) + \alpha_2(\sigma_2^2 + \mu_2^2) + \cdots + \alpha_k(\sigma_k^2 + \mu_k^2)$

Variance of Y is $V(Y) = E(Y^2) - (E(Y)^2) = \{\alpha_1(\sigma_1^2 + \mu_1^2) + \alpha_2(\sigma_2^2 + \mu_2^2) + \cdots + \alpha_k(\sigma_k^2 + \mu_k^2)\} - (\alpha_1\mu_1 + \alpha_2\mu_2 + \cdots + \alpha_k\mu_k)^2$

Now $CV(Y) = \frac{\sigma_Y}{\mu_Y} * 100 = \frac{\sqrt{V(Y)}}{\mu_Y} * 100$

$CV(Y) =$

$$\sqrt{\left\{\alpha_1\left(\frac{\mu_1}{\mu_Y}\right)^2 (CV_1^2 + 100^2) + \cdots + \alpha_k\left(\frac{\mu_k}{\mu_Y}\right)^2 (CV_k^2 + 100^2)\right\} - 100^2}$$

$$= \sqrt{\left\{\omega_1(CV_1^2 + 100^2) + \cdots + \omega_k(CV_k^2 + 100^2)\right\} - 100^2} \tag{1.5}$$

where $\omega_i = \alpha_i(\frac{\mu_i}{\mu_Y})^2$ for $i = 1, 2, \ldots, k$

From equation 1.5, it is observed that $CV(Y)$ can be computed by using α_i, CV_i, and μ_i of X_i, where $i = 1, 2, \ldots, k$. Hence, it is possible to compute $CV(Y)$ without revisiting rudimentary data.

1.4.5.1 Mixture of Normal Distributions

Let Y be the mixture of k univariate normal random variables $X_1, X_2, \ldots, X_k$ with $\alpha_1, \alpha_2, \ldots, \alpha_k$ as mixing parameters. Then, the probability density function (PDF) of Y be $f(x)$ given as in equation (1.6)

$$f(x) = \sum_{i=1}^{k} \alpha_i \frac{1}{\sigma_i \sqrt{2\pi}} e^{-\frac{(x_i - \mu_i)^2}{2}} \tag{1.6}$$

Figure 1.7 shows an example of mixed distribution with $\alpha_1 = 2$ and $\alpha_2 = 1$. The base distributions are f_1 and f_2. f_1 has $\mu = 4$ and $\sigma = 1$, while f_2 has $\mu = 6$ and $\sigma = 2$.

Two normal distributions with means (μ_1, μ_2), standard deviations (σ_1, σ_2) with mixing ratios (α_1, α_2), where $\alpha_1 + \alpha_2 = 1$

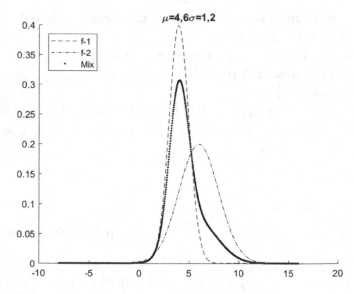

FIGURE 1.7 Mixture distributions example (f_1 has $\mu = 4, \sigma = 1$; f_2 has $\mu = 6, \sigma = 2$).

and $\alpha_1, \alpha_2 \geq 0$ are considered to demonstrate behavior of mixture distribution. Consider the nine cases for $(\mu_1, \mu_2) = \{(4,2), (4,4), (4,6)\}$; $(\sigma_1, \sigma_2) = \{(1,1), (1,0.5), (1,2)\}$ for each mixing ratio $\alpha_1 = 1/2, 1/3$, and $2/3$ (α_2 inferred by $\alpha_1 + \alpha_2 = 1$). Figures 1.8 through 1.10 and Tables 1.1 through 1.3 are for $\alpha_1 = 1/2, 1/3$, and $2/3$, respectively. Here, μ^*, σ^*, and CV^* refer to pooled values with the mixing ratios.

Each subplot of Figures 1.8 through 1.10 provides the PDF of basic two normal distributions with mean and standard deviation as given in the corresponding subplot for mixing α ratios of 1:1, 1:2, and 2:1, respectively. All these subplots follow the same legend as in Figure 1.7. The bimodal behavior can be observed when means are different relative to their respective standard deviations in these plots.

Tables 1.1 through 1.3 for the respective figures provide statistical information (mean, standard deviation, and coefficient of variation) of the base normal PDFs followed by the mixed

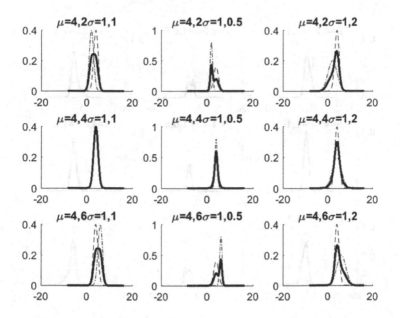

FIGURE 1.8 α Ratio 1:1.

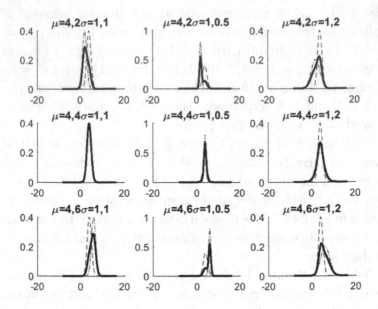

FIGURE 1.9 α Ratio 1:2.

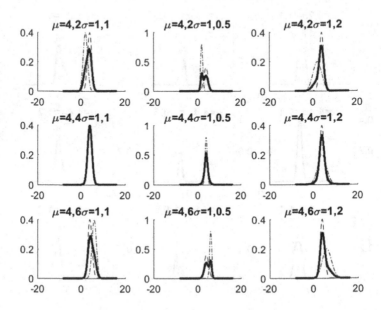

FIGURE 1.10 α Ratio 2:1.

TABLE 1.1 Mixture of 2 Normals with Mixing α Ratio 1:1

S. No.	α_1	α_2	μ_1	μ_2	σ_1	σ_2	CV_1	CV_2	μ^*	σ^*	CV^*
1	0.50	0.50	4	2	1	1	25.00	50.00	3.00	1.41	47.14
2	0.50	0.50	4	2	1	0.5	25.00	25.00	3.00	1.27	42.49
3	0.50	0.50	4	2	1	2	25.00	100.00	3.00	1.87	62.36
4	0.50	0.50	4	4	1	1	25.00	25.00	4.00	1.00	25.00
5	0.50	0.50	4	4	1	0.5	25.00	12.50	4.00	0.79	19.76
6	0.50	0.50	4	4	1	2	25.00	50.00	4.00	1.58	39.53
7	0.50	0.50	4	6	1	1	25.00	16.67	5.00	1.41	28.28
8	0.50	0.50	4	6	1	0.5	25.00	8.33	5.00	1.27	25.50
9	0.50	0.50	4	6	1	2	25.00	33.33	5.00	1.87	37.42

TABLE 1.2   Mixture of 2 Normals with Mixing α Ratio 1:2

S. No.	α_1	α_2	μ_1	μ_2	σ_1	σ_2	CV_1	CV_2	μ^*	σ^*	CV^*
10	0.33	0.67	4	2	1	1	25.00	50.00	2.67	1.37	51.54
11	0.33	0.67	4	2	1	0.5	25.00	25.00	2.67	1.18	44.19
12	0.33	0.67	4	2	1	2	25.00	100.00	2.67	1.97	73.95
13	0.33	0.67	4	4	1	1	25.00	25.00	4.00	1.00	25.00
14	0.33	0.67	4	4	1	0.5	25.00	12.50	4.00	0.71	17.68
15	0.33	0.67	4	4	1	2	25.00	50.00	4.00	1.73	43.30
16	0.33	0.67	4	6	1	1	25.00	16.67	5.33	1.37	25.77
17	0.33	0.67	4	6	1	0.5	25.00	8.33	5.33	1.18	22.10
18	0.33	0.67	4	6	1	2	25.00	33.33	5.33	1.97	36.98

TABLE 1.3   Mixture of 2 Normals with Mixing α Ratio 2:1

S. No.	α_1	α_2	μ_1	μ_2	σ_1	σ_2	CV_1	CV_2	μ^*	σ^*	CV^*
19	0.67	0.33	4	2	1	1	25.00	50.00	3.33	1.37	41.23
20	0.67	0.33	4	2	1	0.5	25.00	25.00	3.33	1.28	38.41
21	0.67	0.33	4	2	1	2	25.00	100.00	3.33	1.70	50.99
22	0.67	0.33	4	4	1	1	25.00	25.00	4.00	1.00	25.00
23	0.67	0.33	4	4	1	0.5	25.00	12.50	4.00	0.87	21.65
24	0.67	0.33	4	4	1	2	25.00	50.00	4.00	1.41	35.36
25	0.67	0.33	4	6	1	1	25.00	16.67	4.67	1.37	29.45
26	0.67	0.33	4	6	1	0.5	25.00	8.33	4.67	1.28	27.43
27	0.67	0.33	4	6	1	2	25.00	33.33	4.67	1.70	36.42

distribution. The first column identifies the mixed distribution for α_1 and α_2 mixing ratios, of base PDFs with means μ_1, μ_2, standard deviations σ_1, σ_2, and coefficient of variations $CV1$, $CV2$, respectively. μ^*, σ^*, and CV^* correspond to the resulting mixture distribution.

Although the base distributions are symmetric, the mixed distribution $f(.)$ needs not be symmetric. The CV of $f(.)$ is dependent on number of its peaks and geometric nature of the peaks. The peaks of $f(.)$ are influenced by base distribution's components and mixing ratios. In particular, the top middle subplots of each Figures 1.8 through 1.10 have CV more than 33% for mixed distribution, though the base components are less than 33%. Rows with identifiers 2, 11, and 20 in Tables 1.1 through 1.3 correspond to these top middle subplots, and the CV of base distributions is consistent in these rows (less than 33%). The geometric properties of peaks and curvatures will help in understanding the feature that has multimodal empirical density function (like relative frequencies) with large CV.

1.5 LIMITATIONS OF COEFFICIENT OF VARIATION

Coefficient of variation is defined for numerical data (real random variable). Coefficient of variation is not definable when mean is zero, it will be unbounded when the mean is approaching zero.

It is recommended to convert all variables to ratio type by appropriate transformation and calibrations [8]. To avoid the non existence of CV, it is recommended to code in strictly positive zone. Further, it is recommended to bring the range of normalization to [1, 2]. So, add constant 1 to the normalization equations used in Section 1.3. Coefficient of variation cannot be defined for multivariate numerical data (multivariate real random variable).

1.6 CV INTERPRETATION

The coefficient of variation is a unit free index. ACV is always nonnegative. In most of the applications, the data will be non-negative, hence CV is equivalent to ACV. In this section, CV is referred without sign. Based on the CV, a system is broadly classified into consistent and inconsistent categories. These are further subdivided into three subcategories based on our exposure and [5]. The following definitions are suggested for abstracting a given system from Machine Learning point of view. Statistical properties of CV estimator are discussed on page 63 of [5] and in [12]. One can refer these for further details to understand the properties of estimator of CV and its comparability with other statistical properties.

Definition 1: *A data (population) will be called as consistent iff $0 < CV < 33$ for that data (population)*

Definition 2: *A data (population) will be called as highly consistent (HC) iff $0 < CV \leq 5$ for that data (population)*

Definition 3: *A data (population) will be called as moderately consistent (MC) iff $5 < CV \leq 15$ for that data (population)*

Definition 4: *A data (population) will be called as weak consistent (WC) iff $15 < CV \leq 33$ for that data (population)*

Definition 5: *A data (population) will be called as weak inconsistent (WI) iff $33 < CV \leq 66$ for that data (population)*

Definition 6: *A data (population) will be called as moderately inconsistent (MI) iff $66 < CV \leq 100$ for that data (population)*

Definition 7: *A data (population) will be called as highly inconsistent (HI) iff $100 < CV$ for that data (population)*

1.7 SUMMARY

These days, the data is acquired mostly by using sensor networks and Internet of Things (IoT) environment. The data are made available for analytics purpose after going through extract transform and load (ETL) process. The data available for analytics is in general a transformed data due to ETL process and due to privacy concerns. One of the classical unified approach of storing the data is normalization, that is the transformed data always lie between 0 and 1. Although CVs of features of original information system are different, the CVs of normalized features (with an assumption that original features are normally distributed) will be identical. Thus, the normalized information system hides some original characteristics.

If a feature has CV larger than 33%, that is, falling under weak to high inconsistency, further investigation is required by decomposing the distribution as a mixed distribution. Refer to [5, 7, 13] for acquiring knowledge of probability and statistics for enhancing decision-making abilities.

1.8 EXERCISES

1. Compute the following features min, max, S_mean, mean, std, acv, ncv, flag , and nflag by considering CV_Threshold as 45, CV_Threshold as 33 for the data sets (a), (b), and (c) given below

- min (minimum of the data)
- max (maximum of the data)
- S_mean (sign of mean of the data)
- mean (magnitude of the mean of the data)
- std (standard deviation of the data)
- acv (absolute coefficient of variation)
- ncv (coefficient of variation of normalized data)
- flag (ACV < CV_Threshold)
- nflag (NCV < NCV_Threshold)

    a. 11, 6, 4, 8, 5, 11, 5, 11, 4, 8, 10, 5, 6, 5, 7, 4, 7, 11, 8, 9, 6, 10, 12, 11, 11, 13, 10, 12, 5, 10, 11, 6, 7, 12, 6, 11, 7, 5, 10, 9, 13, 12, 4, 13, 8, 10, 6, 9, 6

    b. $-3, -4, -9, -11, -9, -5, -7, -9, -7, -7, -7, -7,$ $-8, -3, -11, -7, -12, -8, -11, -12, -12, -10, -3,$ $-12, -12, -6, -10, -8, -6, -10, -7, -9, -10, -3,$ $-11, -8, -5, -5, -11, -6, -11, -3, -9, -5, -3, -5,$ $-6, -4, -12$

    c. 2, −1, 0, −1, −5, −2, −4, 3, 4, 0, −5, 4, −4, 4, 3, −4,
       −4, −3, 4, 2, 4, −5, −3, 3, −4, −1, −2, 0, −2, −4, 1,
       −4, −1, −3, −5, −3, −4, −2, −3, −2, −1, −3, 3, 3, 0,
       −1, 4, −5, −2

2. Compute the feature vector for the data given in exercise 1 by applying translation and scaling.

    a. Translation value $b = 100$ and Scale value $a = 5$

    b. Translation value $b = 200$ and Scale value $a = 50$

    c. Translation value $b = -5$ and Scale value $a = 10$

3. Compute empirical probability of the event

$$X \leq mean \times (1 + p * CV/100)$$

and also empirical probability of the event

$$X \leq mean \times (1 - p * CV/100)$$

for p = 0.1, 0.2, 0.3, 0.4, 0.5, 0.6, 0.7, 0.8, 0.9, 1, 1.25, 1.5, 1.75, 2, 2.5, 3, 4, and 5 for the below given attributes of *Metro Interstate Traffic Volume* Data Set from http://archive. ics.uci.edu/ml [14].

    a. Temp

    b. rain_1h

    c. snow_1h

    d. traffic_volume

4. Derive the Coefficient of Variation for the following distributions

    a. Bernoulli

    b. Binomial

    c. Poisson

    d. Uniform

    e. Beta Type I

    f. Exponential

    g. Normal (Gaussian)

    h. Triangular

5. For the attributes given below from *Metro Interstate Traffic Volume* Data Set available at http://archive.ics.uci.edu/ml [14], compute CV and then generate a decision attribute with CV consistency labels.

    a. Temp

    b. rain_1h

    c. snow_1h

    d. traffic_volume

6. Rank the attributes Temp, rain_1h, snow_1h, and traffic_volume of the dataset *Metro Interstate Traffic Volume* at http://archive.ics.uci.edu/ml [14], based on CV of the attributes.

7. Group the attributes Temp, rain_1h, snow_1h, and traffic_volume based on CV for the dataset *Metro Interstate Traffic Volume* from http://archive.ics.uci.edu/ml [14].

8. Based on consistency type discussed in Section 1.6, group the 45 attributes (excluding ID) of *Parkinson Dataset with replicated acoustic features* available at http://archive.ics.uci.edu/ml [14].

CHAPTER 2

CV Computational Strategies

2.1 INTRODUCTION

The invention of Internet, sensors networks, and IoT facilitated real-time data extraction as precisely as possible, adding features to the data volume, velocity, and variety. The technological growth in the storage systems and high-performance computing gave the opportunity to design the computational strategies to calculate required statistics and then information processing for making decisions to address the real-world problems preferably just in time. The computational methods should sink with various natural topological configurations of data management systems, communication systems, which avoid frequent visits (load, unload) of voluminous raw data. The effectiveness of a recommender system is influenced by all these above factors. One need to design hierarchical and hybrid approaches for building essential required information representation/processing for decision-making process at higher level of abstraction. One of the possible approaches can be designed by exploiting distributed computing paradigm.

The raw data pertaining to local regions can be acquired and maintained at respective local regions by providing a remote access to the data for abstraction purposes. To reduce the data flow, the relevant features for maintaining the data at higher level of abstraction need to be identified to further enhance the throughput. This is possible if one designs Data Mining, pattern recognition, and Machine Learning methodologies on the features, which are possible to derive on the metadata of the respective regional raw data. Such systems demand only metadata flow rather than the raw data. Hence, efficiency of the meta data access, precision, and time will be enhanced. Henceforth, there will be significant speedup of the aggregation process. The CV is one such potential statistical feature that can be derived at higher level of abstractions by using lower statistical measures as metadata. The rest of the chapter provides the computation of CV when data are at a distributed location, which is known as computation of CV for pooled data. The information system or a decision system that is input for any ML system in general has mixed nature. As CV is meaningful to compute only for continuous variables that are of ratio type, there is a need for developing some meaningful transformations such that CV computations can be employed to arrive at meaningful recommendations and apt inference engines.

2.2 CV COMPUTATION OF POOLED DATA

Let X_1, X_2 be two data sets of size n_1 and n_2 at regions R_1 and R_2, respectively. Let X be the *pooled* data from these regions, which is known as *pooled data* of X_1 and X_2 of size $n = n_1 + n_2$ in statistical terms.

Let $n_1, m_1, s_1,$ and CV_1 be the statistical data (number of objects, mean, standard deviation, and CV, respectively) at region1. Similarly, $n_2, m_2, s_2,$ and CV_2 are the statistical data at region2. As s_i is redundant at region$i$ (can be computed using $s_i = \frac{CV_i * m_i}{100}$), the metadata consists of $n_i, m_i,$ and CV_i. The central system pulls this metadata. Central system has two options now:

1. Option1: Compute CV from the pooled data X.

2. Option2: Compute CV from the means and standard deviations of the regions. ($CV = \frac{s}{m} * 100$, where s and m are computed from equations 2.2 and 2.1, respectively.)

Option1 is linear time complex ($O(n)$) task, whereas Option2 is constant time ($O(1)$) task.

To compute CV of pooled data X, mean and standard deviation of X need to be computed.

m, the pooled mean of X, is computed as weighted average of the partitions' means:

$$m = (n_1 * m_1 + n_2 * m_2)/n \qquad (2.1)$$

s, the standard deviation of X, is computed as weighted average of their respective variances

$$s = \sqrt{\frac{1}{n}(n_1 * s_1^2 + n_2 * s_2^2 + n_1 * m_1^2 + n_2 * m_2^2) - m^2} \qquad (2.2)$$

CV, the Coefficient of variation of X is computed as

$$CV = \frac{s}{m} * 100 \qquad (2.3)$$

This computation on two partitions can be extended to k partitions of data. Let i^{th} subcollection of data size be n_i, with mean m_i and coefficient of variation CV_i.

The number of objects is sum of each partition's objects, hence $n = \sum_{i=1}^{k} n_i$.

The pooled mean is weighted average of individual partition's mean. Hence, m, the pooled mean is

$$m = \frac{1}{n} \sum_{i=1}^{k} n_i * m_i \qquad (2.4)$$

The computation of standard deviation of pooled data requires sum of squares. As *Sum of Squares* $= n * (\sigma^2 + \mu^2)$, using ith partition data, sum of squares of ith partition SS_i is computed, as

$$SS_i = n_i * (s_i^2 + m_i^2) = n_i * \left(\left(\frac{CV_i * m_i}{100} \right)^2 + m_i^2 \right) \qquad (2.5)$$

SS, the pooled sum of squares, is

$$SS = \sum_{i=1}^{k} SS_i = \sum_{i=1}^{k} n_i * \left(\left(\frac{CV_i * m_i}{100} \right)^2 + m_i^2 \right) \qquad (2.6)$$

V, the pooled variance, is

$$V = \frac{SS}{n} - m^2 \qquad (2.7)$$

s, the pooled standard deviation, is computed as a square root of the pooled variance.

$$s = \sqrt{V} \qquad (2.8)$$

CV, pooled coefficient of variation, is straightforward to compute now.

$$CV = \frac{s}{m} * 100 \qquad (2.9)$$

2.2.1 Illustration

Consider five samples ($k = 5$) with the following characteristics:

- sample sizes: 75, 32, 74, 35, 85

- sample means: 47.4133, 50.4688, 49.0135, 61.4857, 47.5765

- sample coefficient of variances: 61.4203, 52.2929, 57.8881, 45.3754, 62.9440

Then, the pooled sample size $= 301$.
Pooled sample mean $= 49.8140$ (using equation 2.4).
Pooled coefficient of variance $= 58.3867$ (using equation 2.9).

Algorithm 4 Pooled Coefficient of Variation

Require: k ▷ number of partitions

Require: n_i, m_i, CV_i ▷ number of objects, mean, CV of ith partition

Ensure: m, s, CV ▷ mean, standard deviation and CV of pooled data

1:   $n \leftarrow 0$

2:   $m \leftarrow 0$

3:   $SS \leftarrow 0$

4:   **for** $i \leftarrow 1 \dots k$ **do**

5:      $n \leftarrow n + n_i$

6:      $m \leftarrow m + n_i * m_i$

7:      $SS \leftarrow SS + n_i * ((\frac{CV_i * m_i}{100})^2 + m_i^2)$

8:   **end for**

9:   $m \leftarrow \frac{m}{n}$ ▷ pooled mean

10: $s \leftarrow \sqrt{\frac{SS}{n} - m^2}$ ▷ pooled SD

11: $CV \leftarrow \frac{s}{m} * 100$ ▷ pooled CV

2.3 COMPARISON OF CV WITH ENTROPY AND GINI INDEX

CV can be used in place of entropy and Gini Index for decision-making. Entropy and Gini Index are dependent only on probability of the variable, while CV is dependent on the values of the variable also. In this section, we consider X as a binomial variable in discrete category and normal variable in case of continuous category for discussions.

For a variable with binomial distribution,

$$CV = \frac{\sqrt{n \times p \times (1 - p)}}{n \times p} \times 100 = \sqrt{\frac{(1 - p)}{n \times p}} \times 100.$$

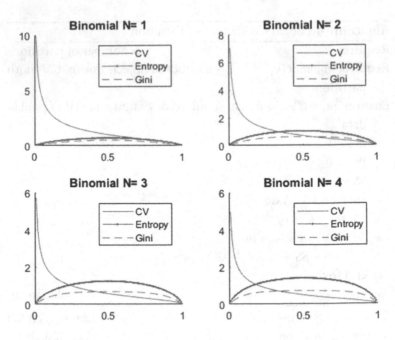

FIGURE 2.1   Binomial distributed variable.

CV is infinity at $p = 0$ and monotonically decreases to zero as p approaches to 1. However, entropy and Gini Index are unimodal and symmetric about $p = 0.5$. Figure 2.1 shows the behavior of CV, entropy, and Gini Index as a function of p (p is along x-axis), p varying from 0 to 1.

For a variable with normal distribution, $CV = \frac{\sigma}{\mu} * 100$. CV monotonically decreases from infinity to 0 as μ goes from 0 to infinity. However, the entropy and Gini Index are constant w.r.t. to μ. Figure 2.2 shows the behavior of CV, entropy, and Gini Index as a function of μ, μ varying from 0 to infinity.

2.4 CV FOR CATEGORICAL VARIABLES

The variables are of four types: (i) nominal, (ii) ordinal, (iii) interval, and (iv) ratio.

The coefficient of variation is meaningful only for ratio type of variables.

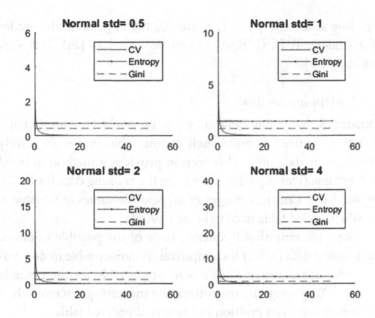

FIGURE 2.2   Normal distributed variable.

The real-word problems consists of mixed data, hence any Machine Learning or data modeling methods should be potential to address the mixed data scenario.

This section provides a systematic method of handling all the kind of data types.

2.4.1 Table Lookup Method

Let X be a categorical variable with possible distinct values, say $a1, a2, \ldots, am$. The project investigator/designer provides coding to these values for the data organization for computerization. Most of the times, the coding is through integers (non-negative). Without loss of generality, we assume the values are coded by integers and the data coding table is provided for data entry/loading. It can be automated by ETL methods using the mapping table provided by the designer.

Illustration: In socio economic scenario, with *income* attribute having values {low, middle, high}, the data encoder can follow a coding scheme = { low:1, middle:2, high:3}. Similarly, grades of a student: F, D, C, B, A, O can be coded as {F:1, D:2, C:3, B:4, A:5, O:6}.

2.4.2 Mapping Method

Sometimes the coding table may not be available from domain, thus there is need to build such coding tables more objectively based on the data only. This section provides a method to build coding table (look-up table). Let D be the training data for building model(s). Consider the set of all possible values of variable X for which coded table need to be generated.

Obtain the empirical frequency table of the possible values of the training data D. Sort the empirical frequency table in decreasing order of frequency. Create a mapping table for the possible values of X by assigning monotonically increasing discrete values corresponding to its position in the sorted ordered table.

Illustration:

Let the attribute Gender in a dataset of 10 tuples have the following values (*Male, Female, Female, Male, Male, Male, Female, Male, Female, Male*). The domain of Gender attribute is *Male, Female* with frequency table

$$Male : 6, Female : 4$$

Hence, we can use the mapping as

$$Male \rightarrow 1, Female \rightarrow 2$$

So the attribute Gender is transformed by this mapping to $\langle 1, 2, 2, 1, 1, 1, 2, 1, 2, 1 \rangle$. Now computation of CV is feasible using these mapped values.

If income group has values *low*, *average*, and *high* frequencies as 30, 60, and 20, respectively, then the coding is {*low:2, middle:1, high:3*}.

If grades of a student are F, D, C, B, A, and O with percentage of students as $F:10\%$, $D:10\%$, $C:20\%$, $B:15\%$, $A:10\%$, and $O:5\%$. Then, the coding scheme is $\{F:6, D:5, C:1, B:2, A:4, O:7\}$.

2.4.3 Zero Avoiding Calibration

As $CV = \frac{\sigma}{\mu} * 100$, to avoid the ill conditioning, it is recommended to convert the values of the variable codes starting from 1, as $1, 2, 3, \ldots$. The codes must not start at 0. This is referred to as *zero avoiding pre-processing* method. Figure 2.3 depicts comparison of CV against entropy and Gini Index for shifted binomial as a probability model for zero avoiding coding. One can notice the CV is unimodal, asymmetric, and skewed towards lower values of p (x-axis in Figure 2.3). This exhibits totally different pattern compared to Figure 2.1, which is a clear indication that CV is sensitive w.r.t. the variable X, while entropy and Gini Index are insensitive (not skewed).

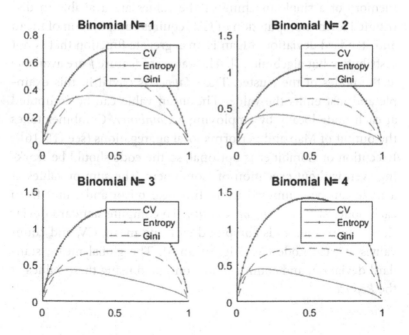

FIGURE 2.3   Binomial distribution with zero avoidance.

2.5 CV COMPUTATION BY MAPREDUCE STRATEGIES

Big Data [15] refers to data having volume, variety, and velocity. Fundamentally, it refers to data that cannot be processed on a single machine (hence distributed), having various types of data like audio, video, text, etc. The data can change with time as in streaming data and social media. Big Data is usually processed using Hadoop [16] environment. It has a distributed file system (called HDFS) and programming model called MapReduce. MapReduce [15, 16] is a programming model for very large data processing. MapReduce is inherently parallel. It consists of two functions: *Map* and *Reduce*. Both of these take key-value pair as input and generate key-value pair.

The MapReduce algorithm shown in Algorithm 5 computes CV of a feature (say f) having a large number of values. This algorithm is to be used when the feature values do not fit in the memory of a single machine or the values are available in distributed form. Computation of CV requires computation of mean and standard deviation. Mean is an aggregate function that is not distributive but algebraic [2]. The values of feature f are available at the nodes of the cluster. The *Mapper* is trivial in this example and just emits the value. The mean value can be computed at each node locally by employing a *Combiner*. Combiner takes the output of Map and performs local aggregations (see [15, 16]). Execution of Combiner is optional, so the code should be working even without execution of Combiners. The sum of values at a node can be computed using the local mean and count from each node, as *sum = mean × count*. To compute standard deviation, sum of squares is computed by using mean, CV, and count values at each node (see Algorithm 4). The global mean, standard deviation, and count can be computed using these values at the *Reducer*.

Algorithm 5 MapReduce Algorithm for CV

1: class **MAPPER**
2: method **MAP**(feature f, value v)
3: **EMIT**(f, $\langle 1, v, v \times v \rangle$) ▷ value is a triplet of count, sum, sum of squares
4:
5: class **COMBINER**
6: method **REDUCE**(feature f, triples $[t1, t2, \ldots]$)
7: $sum \leftarrow 0$
8: $sumSQ \leftarrow 0$
9: $count \leftarrow 0$
10: **for** each value $t \in values[v1, v2, \ldots]$ **do**
11: $\quad sum \leftarrow sum + t.sum$
12: $\quad count \leftarrow count + t.count$
13: $\quad sumSQ \leftarrow sumSQ + t.sumSQ$
14: **end for**
15: $mean \leftarrow sum/count$
16: **EMIT**(f, $\langle$ count, mean, sumSQ $\rangle$) ▷ value is a triplet of count, mean, sum of squares
17:
18: class **REDUCER**
19: method **REDUCE**(feature f, triples $[t1, t2, \ldots]$)
20: $gsum \leftarrow 0$
21: $gcount \leftarrow 0$
22: $gsumSQ \leftarrow 0$
23: **for** each triplet $t \in triples$ **do**
24: $\quad lsum \leftarrow t.mean \times t.count$ ▷ sum of values at each partition
25: $\quad gsum \leftarrow gsum + lsum$
26: $\quad gcount \leftarrow gcount + t.count$
27: $\quad gsumSQ \leftarrow gsumSQ + t.sumSQ$
28: **end for**
29: $gmean \leftarrow gsum/gcount$
30: $variance \leftarrow gsumSQ/gcount - gmean \times gmean$
31: $CV \leftarrow sqrt(variance) * 100/gmean$
32: **EMIT**(f, CV)

2.6 SUMMARY

The need of one-pass algorithms with mixed data analysis is essential to build advisory systems. Building systems with minimal data exchange can handle current Big Data environments having privacy as the prime concern as well as transmission issues of IoT systems. This chapter demonstrated computation of Coefficient of Variation for a distributed data restricting to one pass methodology.

Pre-processing methodologies discussed and demonstrated brought out the adaptability of coefficient of variation for mixed data analysis with scalable computational complexities. The computing strategies discussed in this chapter gel with high-permanence computing (HPC) protocols like pull, push, and virtualization. The hierarchical abstractions of CV demand minimal metadata compared to entropy and Gini Index.

2.7 EXERCISES

1. Compute coefficient of variation and entropy of the empirical distribution for the data of size 80 given in Table 2.1. The corresponding empirical distribution for the data is given in Table 2.2.

TABLE 2.1   Data for Exercise 1

10	11	11	10	10	10	11	12	12	12
7	6	7	7	6	5	6	6	5	7
50	51	50	52	51	51	51	52	52	51
21	21	20	20	21	21	21	22	21	20
11	10	12	12	11	10	12	11	11	12
5	7	5	6	5	6	7	7	6	6
50	51	51	51	52	52	50	52	51	51
21	20	20	20	20	22	21	20	22	21

TABLE 2.2   Empirical Distribution of Data for Exercise 1

Value	5	6	7	10	11	12	20	21	22	50	51	52
Frequency	5	8	7	6	7	7	8	9	3	4	10	6

2. Compute the coefficient of variation for the following data sets and hence the pooled coefficient of variation for the pooled data. Verify the pooled coefficient of variation with that of the pooled data.

  a. Consider four datasets with 15 values each

   i. 11 10 10 10 10 11 10 10 11 10 11 10 11 11 10

   ii. 8 8 6 10 9 6 10 8 6 7 8 6 9 6 5

   iii. 20 22 21 21 22 20 21 22 20 20 20 21 21 21 20

   iv. 12 14 9 9 15 7 13 15 13 7 10 7 11 10 5

  b. Consider four datasets with 15 values each

   i. 10 10 10 12 11 11 11 10 11 11 10 10 11 11 11

   ii. 6 6 6 6 7 6 5 7 6 7 6 7 6 6 6

   iii. 51 50 52 51 51 50 51 50 51 50 52 52 51 51 50

   iv. 20 22 22 21 20 21 21 22 22 20 21 20 22 21 20

3. Obtain the empirical distribution of the categorical data collected from a super market shown in Table 2.3. Provide different kinds of numerical mapping method for the same (Hint: (i) alphabetical order, (ii) simple frequency-based order, (iii) total quantity based order, and (iv) item type and utility based order). Tabulate the corresponding empirical distribution of the numerical coded data. Compare and contrast the empirical distribution of categorical data and its corresponding numerical coded data for each considered mapping.

4. Consider a numerical mapping of *one-to-one nature* and carry out analysis of the categorical data available in Table 2.3, as well as numerically coded data and give your comments.

5. Consider a numerical mapping of *many-to-one nature* and carry out analysis of the categorical data available in Table 2.3, as well as numerically coded data and give your comments.

TABLE 2.3　Super Market Transactions for Exercise 3

S. No.	Grocery Item	Qty	S. No.	Grocery Item	Qty
1	Cloves	11	26	Coriander Powder	4
2	Onions	8	27	Cardamom Powder	3
3	Black Cardamom	6	28	Kismis	1
4	Coriander Powder	4	29	Tamarind	8
5	Kismis	6	30	Cloves	8
6	Cloves	3	31	Green Cardamom	3
7	Green Cardamom	5	32	Kismis	8
8	Green Chilli	5	33	Tamarind	4
9	Cloves	4	34	Cardamom Powder	9
10	Coriander Powder	7	35	Onions	5
11	Cashew Nut	3	36	Kismis	3
12	Coriander Powder	7	37	Green Chilli	10
13	Coriander	5	38	Green Cardamom	7
14	Green Chilli	4	39	Green Cardamom	11
15	Black Cardamom	2	40	Cashew Nut	4
16	Cloves	4	41	Milk	7
17	Milk	9	42	Coriander Powder	8
18	Coriander	8	43	Garlic	3
19	Tamarind	7	44	Cumin Powder	7
20	Milk	9	45	Green Cardamom	8
21	Kismis	9	46	Coriander	8
22	Kismis	9	47	Green Cardamom	6
23	Cinnamon	9	48	Cloves	5
24	Coriander	9	49	Cardamom Powder	10
25	Coriander Powder	7	50	Cinnamon	6

6. Let the coefficient of variation of numerical coded (assume that the mean of numerical coding is positive value) categorical variable is CVC and the entropy of the distribution of the categorical variable be CVE, then show that

a. CVE will be non-negative

b. CVC will be non-negative

c. CVC=0 iff CVE=0

7. Provide a MapReduce computational strategy for computing coefficient of variation by assuming a cluster configuration. Then provide the computational gain factor of the same over serial computation of coefficient of variation. Demonstrate the same by considering small, moderate, and large data set from UCI repository, http://archive.ics.uci.edu/ml [14].

CHAPTER 3

Image Representation

3.1 INTRODUCTION

Image representations are mainly used for content analysis of the image, image recognition, or image classification. These are achieved by derived image representation and feature representation. Some operators are applied on a given image, for example, Sobel operator (see [17]) to arrive at derived image. This derived image is used for detecting edges of an image. Similarly, the three principal components of a color image are obtained and derived image representations are obtained for each principal component. This type of image representation is used for the content analysis of the image, like segmentation, etc. This type of image representation is used to convert one type of image to another, like visual to thermal or visual to X-ray.

Lot of images as part of the video are captured and used for image recognition purpose, in support of cyber investigations. These image recognition systems are developed over the given image databases. Features of a queried image are extracted and similar images are retrieved based on the features, which broadly fall into Content Based Image Retrieval (CBIR). One of the popular CBIR approach is based on seven invariant moments. The seven invariant moments [17] are derived as features of images, organized as an information system for the image database as a representation system. For the query image, the same seven invariant moments are computed and used as a feature vector. This feature vector is used to retrieve tuples and corresponding images.

This chapter focuses on using CV as an operator for a window of size w to arrive at a derived image (CVImage). The CV features (discussed in Section 1.4.4) and bit vectors corresponding to predefined CVThresholds for a given image are used as feature vector (referred as CVFeatures of image).

3.2 CVIMAGE

Image is a two-dimensional array of pixels where each pixel is represented by RGB vector (three dimensional) for color image and single dimensional for grayscale image.

CVImage is the representation of an image by using coefficient of variation. For a given image X, *CVImage* is obtained by replacing every pixel of X with CV of a window of size $w \times w$. Let us denote it by $CVI(X)$. $CVI(X)$ is formed by replacing every $(i,j)^{\text{th}}$ pixel of the image by CV of the window of size $w \times w$ centered at pixel (i,j). The CV values are rounded to integer to fit the image format. When an image X is of size $n \times m$, then the Algorithm 6 returns $CVI(X)$, which is of size $(n-w+1) \times (m-w+1)$. *CVImage* of size $n \times m$ can be obtained by appropriate zero padding to the original image.

An image corresponding to edges of the given image is obtained by arriving at a binary representation of *CVI*. Let *CVT* be the threshold on CV of $(i,j)^{\text{th}}$ pixel with the proposed windowing method. A binary representation of the image is obtained by the test $CVI(i,j) > CVT$. Each pixel can be either 1 or 0 based on this test. Further $CVIF(X)$ is indicator matrix of *CVI* based on the *CVT*. This is applicable for both grayscale and color images.

Algorithm 6 CVImage

Require: I, w, CVT ▷ Image I, window w, CVThreshold CVT
Ensure: $CVI, CVIF$
1: $X \leftarrow readImage(I)$
2: $n \leftarrow rows(X)$
3: $m \leftarrow cols(X)$
4: $k \leftarrow$ dimension of pixels of X
5: **for** $i \leftarrow 1$ to $n - w$ **do**
6:     **for** $j \leftarrow 1$ to $m - w$ **do**
7:         $XW \leftarrow$ array of pixel values of X using window of size $w \times w$ with (i, j) at center
8:         **for** $l \leftarrow 1$ to k **do**
9:             $CVI(i, j, l) \leftarrow CV(XW)$
10:             $CVIF(i, j, l) \leftarrow CV(XW) > CVT$
11:         **end for**
12:     **end for**
13: **end for**

The choice of CVT can be decided by using consistency levels of CV as defined in Section 1.6.

Figure 3.1 demonstrates the computation of a step of CVI and CVIF. CV of the nine values in the dotted square of size 3×3 in sub figure (a) is shown as the pixel value at the center in sub figure (b) in bold. The corresponding flag value is shown in center of the square of sub figure (c) by taking CVT = 33.

Figure 3.2 illustrates CVImage for a sample image by using CVT = 33% and w = 3.

3.2.1 CVImage Representation of Lena Image

Figure 3.3 provides an illustration of CVImage representation of Lena (Figure 3.3a) [11]. It depicts Lena grayscale image and its moving average image (known as a smoothed image) with window size of W = 3, the corresponding images of standard

2	6	5	1	4	1
2	7	3	1	2	3
9	2	6	2	5	1
3	1	1	2	8	4
1	0	7	3	2	6
2	3	5	1	1	3

(a)

2	6	5	1	4	1
2	52	3	1	2	3
9	2	6	2	5	1
3	1	1	2	8	4
1	0	7	3	2	6
2	3	5	1	1	3

(b)

2	6	5	1	4	1
2	1	3	1	2	3
9	2	6	2	5	1
3	1	1	2	8	4
1	0	7	3	2	6
2	3	5	1	1	3

(c)

FIGURE 3.1 CVImage illustration: (a) Pixels values, (b) CVImage demo, and (c) CVIF demo.

deviation, and coefficient of variation. The rest of Figure 3.3b–d depict respective binarized images [17] of Lena considering CVT as 3, 9, . . . , 63. Figure 3.3b shows binarized images of Lena with CVT as 3, 9, 15, and 21, respectively. Figure 3.3c shows binarized images of Lena with CVT as 27, 33, 39, and 45, respectively. Figure 3.3d shows binarized images of Lena with CVT as 51, 57, 63, and 69, respectively.

Figure 3.3b to shows the various binarized images as CVT increases monotonically. Hence, the number of nontrivial pixels decrease monotonically. It can be observed that low CVT captures **texture**, medium values of CVT capture **edges**, while high

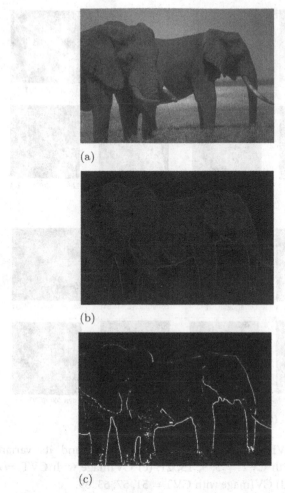

(a)

(b)

(c)

FIGURE 3.2 CVImage: (a) Sample image, (b) CVImage of sample image, and (c) CVImage with CVT = 33% for edge representation.

CVT captures **outlines**. For this Lena image, low CVT = 3 to 21, medium CVT = 27 to 45, and high CVT = 51 to 69. So, one can adopt different CVT values for adaptable binarization of an image. This type of visualization assists in understanding the given system with various perceptions.

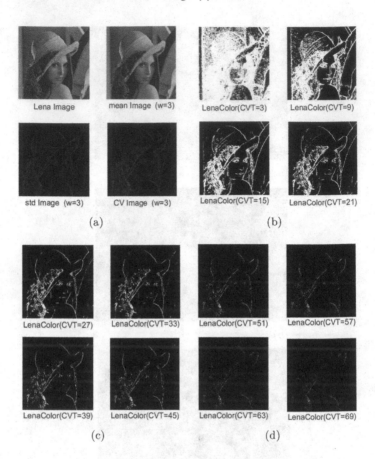

FIGURE 3.3 CVImage of Lena: (a) Lena image and its variants; (b) CVImage with CVT = 3, 9, 15, 21; (c) CVImage with CVT = 27, 33, 39, 45; and (d) CVImage with CVT = 51, 57, 63, 69.

3.3 CV FEATURE VECTOR

Considering the pixel values as the data, statistical features are computed. In case of color images, pixel has three values. For each color plane data, the corresponding statistical features (see [8]) will be computed. The histogram of the data encapsulates all statistical information. From the histogram, one can compute the statistical features like mean, mode, median, inter quartile range, SD, CV, entropy, and so on. For a given color image, one can extract CV for each color plane, viz., CVR, CVG, CVB as a three-dimensional real feature vector. For a given threshold (as recommended in Section 1.6), a bit vector of size 3 can be derived for the image. If one considers a set of thresholds, then that many bit vectors can be derived. If k distinct thresholds are considered, there will be $3k$ bit vectors. Considering the gray image of the color image, one can get coefficient of variation of CVGray, and it induces k bit vectors for the corresponding k thresholds. Thus, four-dimensional CV vectors and $4k$ bit vectors for k thresholds can be treated as CV feature vector. One can restrict to the subset for representing the image for their application.

Figure 3.4 depicts a color image and the histograms of Red Channel, Green Channel, and Blue Channel in subfigures (Computation, Representation and Analysis are carried out on color image though the figures of this book are printed in grey color). The support of pixel values corresponding to Red (R) vary from 0 to 255, corresponding to Green (G) vary from 0 to 200, and corresponding to Blue (B) vary from 0 to 100 (approximate values).

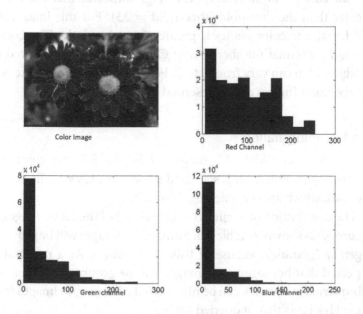

FIGURE 3.4   Color image and histograms of its color channels.

TABLE 3.1   Image Statistics

Color	Mean	Standard Deviation	CV	CV flag
Red	96.23875	63.82992	66.32455	1
Green	47.23037	49.63251	105.086	1
Blue	21.55518	27.66961	128.3664	1

The statistical values corresponding to this image are given in Table 3.1 with a single threshold of CVT = 33%. Thus, the CV feature vector for this image is CVR, CVG, CVB, and corresponding flags with threshold = 33%. The feature vector for this image is CVR = 66.32, CVG = 105.09, CVB = 128.37, CVRF = 1, CVGF = 1, and CVBF = 1. We define the corresponding decimal value of the flag vector as a Hvalue (*Hvalue*). For this image, Hvalue = 7. Embedding Hvalue as one of the features, CV feature vector for this image is seven dimensional. Hence, the feature is CVR = 66.32, CVG = 105.09, CVB = 128.37, CVRF = 1, CVGF = 1 and CVBF = 1, Hvalue = 7.

Last column of Table 3.1, "CV flag," indicates that the CV is greater than the threshold (threshold = 33). For this image, the CV for all the color planes is greater than threshold. The corresponding decimal number (using CVflag of RGB) is referred as Hvalue which can vary from 0 to 7. For this image, the Hvalue is 7. Hence, color images are represented using Hvalues.

3.3.1 Demonstration

This section considers 20 images of Figure 3.5 and computes the corresponding means, standard deviations, CVs, CVflags, and Hvalues, which are compiled in Table 3.2.

The distribution of Hvalues (Hash values) of the set of images of Figure 3.5 is shown in Table 3.3. Similar CV images will be grouped together (granule) because of this Hash values. As a result, it is expected that homogeneous images will be grouped together. So, this queried image task can be minimized confining to images with same Hvalue as that of queried image.

Imagld:1   Imagld:2   Imagld:3   Imagld:4

Imagld:5   Imagld:6   Imagld:7   Imagld:8

Imagld:9   Imagld:10   Imagld:11   Imagld:12

Imagld:13   Imagld:14   Imagld:15   Imagld:16

Imagld:17   Imagld:18   Imagld:19   Imagld:20

FIGURE 3.5   SampleDB.

3.3.2 Hvalue Distribution Analysis

CV features (mean, standard deviation, CV, CVflag, Hvalue) of eight image databases with the image details as given in Table 3.4, are given in Appendix tables, with one table for every image database. These databases' Hvalue distribution is given in Table 3.5. Inferences on these databases can be drawn from this distribution of Hvalues. For instance, the Hvalues distribution of "Greenlake" database of 48 size has turned out to be point distribution as probability(Hvalue = 7) is 1. While the distribution of Hvalue for "Indonesia" database is probability

TABLE 3.2   Hash Values

ID	μ_R	σ_R	CVR	CVRF	μ_G	σ_G	CVG	CVGF	μ_B	σ_B	CVB	CVBF	Hvalue
1	188.13	28.23	15.00	0	135.49	31.16	23.00	0	85.80	34.20	39.86	1	1
2	192.57	53.90	27.99	0	135.36	38.64	28.54	0	103.89	38.49	37.05	1	1
3	194.92	68.76	35.28	1	173.59	53.93	31.07	0	154.54	61.95	40.09	1	5
4	96.24	63.83	66.32	1	47.23	49.63	105.09	1	21.56	27.67	128.37	1	7
5	114.11	69.56	60.96	1	118.72	56.33	47.45	1	50.28	40.35	80.26	1	7
6	136.29	52.30	38.37	1	161.06	62.16	38.59	0	188.27	63.95	33.97	1	7
7	99.03	18.13	18.31	0	124.48	18.20	14.62	0	171.01	21.48	12.56	0	0
8	73.01	57.24	78.40	1	76.97	38.48	50.00	1	51.86	27.42	52.87	1	7
9	191.09	53.31	27.90	0	172.62	62.42	36.16	1	191.33	64.47	33.69	1	3
10	123.23	63.50	51.53	1	102.37	66.31	64.78	1	84.45	57.04	67.54	1	7
11	93.90	28.71	30.57	0	99.83	31.73	31.78	0	83.60	40.23	48.12	1	1
12	138.41	35.98	26.00	0	137.43	34.99	25.46	0	116.00	46.01	39.66	1	1
13	87.55	29.60	33.81	1	94.48	30.84	32.64	0	96.47	22.97	23.81	0	4
14	100.63	34.90	34.68	1	105.40	39.25	37.24	1	107.92	29.47	27.31	0	6
15	65.73	27.95	42.52	1	73.29	21.68	29.58	0	17.63	21.66	122.86	1	5
16	132.61	36.67	27.65	0	115.96	33.99	29.31	0	77.45	26.06	33.65	1	7
17	134.93	95.31	70.64	1	96.56	84.58	87.59	1	72.87	63.08	86.56	1	7
18	101.35	40.73	40.18	1	98.25	42.79	43.55	1	90.16	50.58	56.10	1	7
19	137.74	45.87	33.30	1	114.23	36.76	32.18	0	100.46	35.69	35.53	1	5
20	90.95	29.16	32.06	0	111.43	40.81	36.63	1	118.87	64.10	53.92	1	3

TABLE 3.3   Hash Value Distribution of Sample DB

Hash Value	Frequency	Empirical Probability	Cumulative Probability Distribution
0	1	0.05	0.05
1	6	0.3	0.35
2	0	0	0.35
3	2	0.1	0.45
4	1	0.05	0.5
5	2	0.1	0.6
6	1	0.05	0.65
7	7	0.35	1
Total	20		

TABLE 3.4   Image Databases

ID	DB Name	#Images	Type of Images
1	Greenlake	48	Lake images from [18]
2	Greenland	256	Outdoor images from [19] converted to Gray scale
3	TrainDB	200	Heterogeneous collection of images
4	TestDB	100	Heterogeneous collection of images
5	Indonesia	36	Outdoor Landscapes and structures' images from [20]
6	SampleDB	20	Heterogeneous collection of images
7	Testimages	33	Outdoor images
8	TextureDB	4	Texture
	Total	697	

(Hvalue $= 1$) $= 0.11$ and probability(Hvalue $= 7$) $= 0.89$. The values indicate that all the images in "Greenlake" database are homogeneous with high CVs; this happened so that all the images are outdoor images for "Greenlake" database. While "Indonesia" database is not totally homogeneous, though all are outdoor images. Some images of this database have high CV only for Blue color plane. This additional knowledge is possible due this representation.

TABLE 3.5   Hvalue Distribution Database Wise

Image DB Name	DB Id	Hash Value							Grand Total
		0	1	3	4	5	6	7	
Greenlake	1					1		47	48
Greenland	2		256						256
TrainDB	3	18	13	3	4	9	4	149	200
TestDB	4	8	4	2		6		80	100
Indonesia	5		4					32	36
SampleDB	6	1	6	2	1	2	1	7	20
Testimages	7					1		32	33
Images_Texture	8	1				1		2	4
Grand Total		28	283	7	5	20	5	349	697

Based on this empirical study, the chances of Hvalue for an arbitrary image being 7 is 0.5, then Hvalue of 1 with 0.4, covering around 90% of probability. It is observed that Hvalue = 2 is a trivial event. Inconsistency behavior only in Green plane seems impossible in these databases.

3.3.3 Ranking of CV Features

This section demonstrates the importance of CV features discussed in Section 3.2 for image representation. The feature ordering is one of the useful techniques for Data Mining and data analytics. One can employ existing methodologies under supervised and unsupervised methods [2, 21]. In this section, for the considered images of Table 3.4, Random Forest method [2] is adopted under supervised category considering Hvalue as class label (dependent variable). The features in the ranking order are. CVR, cvfR, CVG, cvfG, CVB, cvfB, sdG, sdR, sdB, CVGray, CVGrayF, meanB, meanG, sdGray, meanR, and meanGray.

For each attribute, the CV has been considered as a metric under unsupervised ranking approach. The order of the features is meanGray, CVGray, sdGray, CVGrayF, meanG, meanR, cvfB, sdG, sdR, meanB, sdB, CVB, CVG, cvfR, CVR, and cvfG. When only colored images are considered, the ranking order is meanG, meanR, cvfB, sdG, sdR, meanB, sdB, CVB, CVG, cvfR, CVR, and cvfG.

In unsupervised methodology, the feature mean represents smoothed image of the original, retaining absolute information. This automatically inherits inherent variations. Thus, mean vectors received better ranking over the SD and CV. When in supervised learning, the Hvalue is derived based on the CV discretized value, thus CV receives better ranking over other features. Which is consistent with definition and derivation of Hvalue.

3.4 SUMMARY

The potential knowledge representation characteristics of CVImage are demonstrated for classical Lena image.

The CVThreshold can be fixed depending on the application requirement, which facilitates adaptability of binarization of a given image.

CVFeatures of images are derived and demonstrated for benchmark image databases available at [10].

The choice of number of distinct CV thresholds will give rise to bit vector of size 3 times of the distinct thresholds for color image (ignoring gray). The corresponding decimal value give rise to Hvalue $0, 1, \ldots, 2^{3*k-1}$. Increasing k leads to finer granular representation, hence, reduction in search complexity. This feature representation can be enhanced by considering image in different forms like RGB, HSV. The corresponding CV features can be composed as a feature vector of the given image. One can investigate merits and limitations of these feature vectors.

3.5 EXERCISES

1. Convert the Lena color RGB image into HSV format (hue, saturation, value [17]), compute the Hvalue for the HSV image.

2. Find out the distribution of empirical distribution of Hvalues by converting into HSV for the images given in Figure 3.5.

3. Construct a feature vector FV by considering coefficient of variation of each color plane of a given color image. Carry out a cluster analysis on the training data build by FVs for the images given in Figure 3.5. Derive the corresponding cross-tabulation of cluster id and Hvalue and carry out contingency analysis on the same.

4. Consider the feature vector FV=[CVGray, CVR, CVG, CVB] from the tables of Appendix to carry out cluster analysis. Consider the cluster id as a new categorical variable and append to the table. Carry out the contingency analysis between cluster id and Hvalue and give your comments.

5. Consider the feature vector FV=[CVGray, CVR, CVG, CVB, Hvalue] from the tables of Appendix as a decision system with Hvalue as decision variable, perform classification using the methods:

   a. CART
   b. ID3
   c. k-NN

6. Generate CVImage for the image data bases: *iran, australia* and *football* available at http://imagedatabase.cs.washington. edu/groundtruth/.

7. Obtain the CV feature representation for the image databases given in Exercise 6.

8. Obtain the empirical distribution of Hvalues for the image databases of the Exercise 6 and give your observations and comments.

CHAPTER 4

Supervised Learning

4.1 INTRODUCTION

Supervised learning refers to training a machine with labeled data to predict correct outcome for unseen data. The data used to train the model (called as *train data*) consists of correct label for every data object. The model performance is evaluated on *test data* (which is not seen by the model during training phase). Classification and regression are two major approaches for supervised learning. Coefficient of variation (CV) can be used to develop models for both of these approaches.

This chapter develops and demonstrates the potential characteristics of CV for building supervised learning models. It introduces CV Gain in lines of Information Gain for attribute selection and demonstrates decision tree/regression tree induction. The preprocessing approaches required for CVGain computation and the merits of CVGain are described in this chapter. The merits of CVGain computation makes it highly suitable for Big Data environment as well as applications where the decision can be fuzzy.

The train data consists of conditional attributes or features used to predict the outcome. Conditional attributes are independent variables that will be used to predict the dependent variable. Let the data set $\mathcal{D}$ consists of features $X_1, X_2, \ldots, X_n$ and a dependent variable Y. Let each feature X_i be from domain D_i and Y has domain D_Y. Then $\mathcal{D} = \{(\boldsymbol{x}_i, y_i) \mid \boldsymbol{x}_i \in D_1 \times D_2 \times \cdots \times D_n, y_i \in D_Y\}$

where data object x_i has output y_i. For classification, D_Y is categorical and for regression, D_Y is numeric. The CV of the dependent variable Y is given by equation 4.1,

$$CV(Y) = \frac{\sigma(Y)}{\mu(Y)} \times 100 \tag{4.1}$$

4.2 PREPROCESSING (DECISION ATTRIBUTE CALIBRATION)

An example train data is given in Table 4.1, having four features (Outlook, Temperature, Humidity, and Wind) and a decision attribute for Play (source [22]). This data set has a categorical dependent variable, and hence is suitable for classification. CV computation requires mean and standard deviation, but these cannot be computed for categorical data. To facilitate CV computation, the categorical attributes need to be transformed using the mapping method described in Section 2.4.2. Considering the decision attribute *Play*, the mapping is *Yes* → 1 and *No* → 2 as

TABLE 4.1 Train Data for Play Tennis

Day	Outlook	Temperature	Humidity	Wind	Play
1	Sunny	Hot	High	Weak	No
2	Sunny	Hot	High	Strong	No
3	Overcast	Hot	High	Weak	Yes
4	Rain	Mild	High	Weak	Yes
5	Rain	Cool	Normal	Weak	Yes
6	Rain	Cool	Normal	Strong	No
7	Overcast	Cool	Normal	Strong	Yes
8	Sunny	Mild	High	Weak	No
9	Sunny	Cool	Normal	Weak	Yes
10	Rain	Mild	Normal	Weak	Yes
11	Sunny	Mild	Normal	Strong	Yes
12	Overcast	Mild	High	Strong	Yes
13	Overcast	Hot	Normal	Weak	Yes
14	Rain	Mild	High	Strong	No

the frequency information is *Yes*:9 and *No*:5. The feature *Outlook*'s frequency information is *Sunny*:5, *Rain*:5, and *Overcast*:4. So, the mapping can be *Sunny* → 1, *Rain* → 2, and *Overcast* → 3. This transformation is required only for decision attribute as will be observed soon in further sections. After this mapping, the dependent variable *Play* has the values 2, 2, 1, 1, 1, 2, 1, 2, 1, 1, 1, 1, 1, 2; hence, *mean* = 1.36, *standard deviation* = 0.48, and *CV* = 35.31.

CV computation requires *nonzero mean* (Section 2.3). If *mean* is near to zero, CV is sensitive for small changes in *mean*. Hence, the mapping procedure is chosen such that zero is avoided (see Section 2.4.3).

4.3 CONDITIONAL CV

CvGain of an attribute is computed with the help of its Conditional CV. Conditional CV explains the dispersion of the decision attribute with respect to an independent attribute. Conditional CV is computed using only those records in which the feature has a particular value. Conditional CV of an attribute Y is computed by using a subset of its values. CV of Y conditioned on attribute X is denoted as $CV(Y|X)$. It serves as a measure of dispersion in Y with respect to the values of X. For example, $CV(Play|Outlook)$ is computed by considering *Play* values for *Outlook* values {*Sunny*, *Overcast*, *Rain*}.

Conditional CV of Y w.r.t. feature A_i having value x, i.e., $CV(Y|A_i = x)$ is computed using the records of DT where the ith feature has value x, i.e., $A_i = x$. Conditional CV of Y w.r.t. a feature A_i, when A_i has v distinct values $a_1, a_2, \ldots, a_v$ is computed as weighted average using probabilities for weights. This computation is given by equation 4.2. $P(A_i = a_j)$ is the probability of A_i having value a_j.

$$CV(Y|A_i) = \sum_{j=1}^{v} P(A_i = a_j) \times CV(Y|A_i = a_j) \qquad (4.2)$$

Consider the decision system with the decision attribute *Play* having mapped values as shown in Table 4.2. The features *Outlook*,

TABLE 4.2   Play Tennis Data

Day	Outlook	Temperature	Humidity	Wind	Play
1	Sunny	Hot	High	Weak	2
2	Sunny	Hot	High	Strong	2
3	Overcast	Hot	High	Weak	1
4	Rain	Mild	High	Weak	1
5	Rain	Cool	Normal	Weak	1
6	Rain	Cool	Normal	Strong	2
7	Overcast	Cool	Normal	Strong	1
8	Sunny	Mild	High	Weak	2
9	Sunny	Cool	Normal	Weak	1
10	Rain	Mild	Normal	Weak	1
11	Sunny	Mild	Normal	Strong	1
12	Overcast	Mild	High	Strong	1
13	Overcast	Hot	Normal	Weak	1
14	Rain	Mild	High	Strong	2

Temperature, Humidity, and *Wind* determine the dependent attribute, *Play,* having two class labels 1 and 2 corresponding to <*Yes, No*>. The *CV* of attribute *Play* is $\sigma(Play)/\mu(Play) \times 100 = 36.63$. The *conditional CV* of *Play* w.r.t. *Outlook* is:

$$CV(Play|Outlook) = \frac{5}{14} \times CV(Play|Outlook = Sunny)$$

$$+ \frac{4}{14} \times CV(Play|Outlook = Overcast)$$

$$+ \frac{5}{14} \times CV(Play|Outlook = Rain)$$

$CV(Play|Outlook = Sunny)$ is computed by considering the class labels of *Play* when *Outlook = Sunny*. The values of *Play* when *Outlook = Sunny* are 2, 2, 2, 1, and 1. *CV* of these values is 30.61. With similar computations for *Outlook = Overcast, CV* is zero, as all the four values belong to the same class; for *Outlook = Rain, CV = 34.99* (the values of *Play* when *Outlook = Rain* are 1, 1, 2, 1, and 2). Hence,

$CV(Play|Outlook) = \frac{5}{14} \times 30.61 + \frac{4}{14} \times 0 + \frac{5}{14} \times 34.99 = 23.43.$

4.4 CVGAIN (CV FOR ATTRIBUTE SELECTION)

CV is a measure of dispersion or variation of an attribute. It is used for meaningful comparisons of two or more magnitudes of variation, even if they have different *means* or different scales of measurement. CVGain of an independent attribute quantifies the reduction in variance of a dependent attribute w.r.t. the independent attribute.

CVGain of a feature is defined as the difference of *CV* of decision attribute Y and its *conditional CV*. It depicts how much variance is reduced in Y by using the feature. The formula for *CVGain* is as given in equation 4.3. *CVGain* of a feature is an estimate of its ability to split the data set records into homogeneous groups. The attribute with highest *CVGain* will reduce the variation the most.

$$CVGain(Ai) = CV(Y) - CV(Y|A_i) \qquad (4.3)$$

4.4.1 Example

Considering the data of Table 4.1, $CVGain(Outlook) = CV(Play) - CV(Play|Outlook)$. From Section 4.2 $CV(Play) = 35.31$ and from Section 4.3 $CV(Play|Outlook) = 23.43$. Hence, *CVGain* $(Outlook) = 35.31 - 23.43 = 11.88$.

Similarly, *CVGain* of remaining attributes (*Temperature, Humidity,* and *Wind*) can be computed.

4.5 ATTRIBUTE ORDERING WITH CVGAIN

Independent variables of an information system can be ordered using CVGain. The attributes with high CVGain are more important than those with low CVGain.

From previous section, CVGain(Outlook) = 11.88. CVGain of the *Temperature* attribute is computed for its values *Hot, Mild,* and *Cool.* There are four records with *Temperature = Hot,* having Play = 2, 2, 1, and 1. Therefore, $CV(Play|Temperature = Hot) = 33.33$. With the six records for *Temperature = Mild,* and

Play $= 1, 2, 1, 1, 1,$ and 2; $CV(Play|Temperature = Mild) = 35.35$. From $Temperature = Cool$ and Play $= 1, 2, 1,$ and 1; $CV(Play|Temperature = Cool) = 34.64$.

Hence,

$$CV(Play|Temperature)$$
$$= \frac{4}{14} \times 33.33 + \frac{6}{14} \times 3.35 + \frac{4}{14} \times 34.64 = 34.57$$

Therefore,

$$CVGain(Temperature) = CV(Play)$$
$$- CV(Play|Temperature)$$
$$= 35.30 - 34.57 = 0.73$$

Similarly,

$$CVGain(Humidity) = CV(Play) - CV(Play|Humidity)$$
$$= 35.30 - \left(\frac{7}{14} \times 31.49 + \frac{7}{14} \times 30.61 \right)$$
$$= 4.25$$
$$CVGain(Wind) = CV(Play) - CV(Play|Wind)$$
$$= 35.30 - \left(\frac{8}{14} \times 34.64 + \frac{6}{14} \times 33.33 \right)$$
$$= 1.22$$

Now the ordering of attributes using CVGain (in descending order) is

- *Outlook* (CVGain $= 11.88$)
- *Humidity* (CVGain $= 4.25$)
- *Wind* (CVGain $= 1.22$)
- *Temperature* (CVGain $= 0.73$)

4.6 CVDT FOR CLASSIFICATION

Decision tree is a popular supervised learning technique to perform classification. Few merits of decision trees are simplicity, interpretability, and feature selection. Many types of decision trees are available like ID3, C4.5, CART, etc. These decision trees mainly differ in the attribute selection criteria like Information Gain (based on entropy) and Gini Index to split the initial set of data objects into homogeneous subgroups. Decision tree has decision nodes as internal nodes and leaf nodes with class labels. Decision nodes contain test condition on independent attributes. Decision tree induction procedure requires choosing an attribute for splitting the data into homogeneous groups w.r.t. decision attribute. See [2, 3, 22] for detailed discussion of decision trees.

CVGain is the measure to decide which feature (or independent variable) of the data will give the best split of the initial set of data objects into homogeneous subsets of data (which can decrease the variance in child node w.r.t. parent node the most).We can use the attribute ordering shown in Section 4.5 for selecting the attribute for splitting the objects at decision nodes. So, decision tree constructed using CVGain is named CV-based Decision Tree (*CVDT*) [23].

Decision tree is built by greedy approach of recursively selecting the best attribute to reduce variance. The popular Hunt's algorithm [2, 3, 22] is used with *CVGain* as attribute selection measure. This algorithm repeatedly partitions the data by the attribute with highest *CVGain*. The attribute that reduces the variance the most according to the class attribute is identified by *CVGain*. If the splitting attribute has multiple values, then the decision tree has multiway splits. This process is repeated for every subtree resulting from the split and stopped when it is not possible to split further (all data tuples belong to the same class or no more attributes to split).

Unlike Information Gain, CVGain can be negative. If CVGain is negative for an attribute, that attribute should not be used for attribute selection measure. Entropy and Gini Index depend on frequency (Probability) distribution but not on attribute values.

CV depends on attribute values as well as frequencies. For example, if attribute X is Bernoulli with P and Q as probabilities, the Entropy and Gini Index can be computed, while X values are also required for computing CV. Let X_1 be Bernoulli with possible values 5 and 10 with probabilities P and Q. Let X_2 be Bernoulli with possible values 25 and 50 with probabilities P and Q. The Entropy of random variables X_1 and X_2 will be same and also same is the case with Gini Index. But CV will differ for X_1 and X_2, indicating potential characteristics of CV to distinguish distinct objects. CVDT can handle categorical and numeric attributes (see preprocessing of categorical attributes given in Section 4.2). The conventional pruning, scalable, and ensemble approaches of decision trees are applicable for CVDT as well.

4.6.1 CVDT Algorithm

The procedure for constructing the CVDT from training data is shown in Algorithm 7. The tree is constructed in top-down recursive manner. Initially, all the training data is at the root node. Lines 7–9 are for terminating the recursion when all records belong to the same class. Lines 10–12 will terminate the recursion for independent attributes list being empty, using majority voting to label with the most frequent class in that data. Line 14 finds the attribute with highest CVGain. Lines 15–22 will grow a branch from the current node and attach the CVDT obtained for each partition based on splitting attribute.

4.6.2 CVDT Example

The training data for building the CVDT is taken from Table 4.2. The class attribute is *Play*, having two labels, 1 and 2 (1 for Yes and 2 for No). The independent attributes, *Outlook*, *Temperature*, *Humidity*, and *Wind* are all categorical. The CV of attribute *Play* is 36.63 (its standard deviation is 0.4791 and mean is 1.371). *CVGain* of attribute *Outlook* is **11.87** from Section 4.4.1.

As *Outlook* has highest CVGain (see Section 4.5 for attribute ordering by CVGain), *Outlook* is selected as the splitting attribute.

Algorithm 7 CVDT Induction

Require: $DT, D, A = \{A_1, A_2, \ldots A_n\}$ ▷ D is the decision attribute, A is the set of features of Decision Table DT

1:   $N \leftarrow createNode(DT)$
2:   **if** $CV(D) == 0$ **then**
3:      $label(N) \leftarrow C$ ▷ C is the class of all tuples of DT
4:      return N as leaf
5:   **end if**
6:   **if** A is empty **then**
7:      $label(N) \leftarrow C$ ▷ C is the majority class in DT
8:      return N as leaf
9:   **end if**
10:   $A_s \leftarrow \arg\max CVGain(DT, A, D)$ ▷ A_s is the splitting attribute
11:   $A \leftarrow A - A_s$ ▷ A_s is categorical
12:   **for all** $v_j \in A_s$ **do**   ▷ partition the tuples and grow sub trees
13:      $DT_j \leftarrow$ Partition of DT with $A_s = v_j$
14:      **if** DT_j is empty **then**
15:          attach leaf node to N, labeled with majority class in DT
16:      **else**
17:          attach tree $CVDT(DT_j, A, D)$ to N
18:      **end if**
19:   **end for**

Using this as splitting attribute, the root node having data of Table 4.2 is divided into three tables forming the child nodes of root node of decision tree as shown in Tables 4.3 through 4.5. Three branches are grown from the root node, labeled *Outlook = Sunny, Outlook = Overcast,* and *Outlook = Rain*.

Now using the data from Table 4.3 (node along the branch *Outlook = Sunny*), we have to compute CVGain for *Temperature, Humidity,* and *Wind.* Out of these attributes, *Humidity* has the highest CVGain of 30.61. Further partition on *Humidity* generates two leaf nodes, along the branch *Humidity = High*, class label = 2 (No); along the branch *Humidity = Normal*, class label = 1 (Yes).

TABLE 4.3   Decision Table for Outlook = Sunny

Temperature	Humidity	Wind	Play
Hot	High	Weak	2
Hot	High	Strong	2
Mild	High	Weak	2
Cool	Normal	Weak	1
Mild	Normal	Strong	1

TABLE 4.4   Decision Table for Outlook = Overcast

Temperature	Humidity	Wind	Play
Hot	High	Weak	1
Cool	Normal	Strong	1
Mild	High	Strong	1
Hot	Normal	Weak	1

TABLE 4.5   Decision Table for Outlook = Rain

Temperature	Humidity	Wind	Play
Mild	High	Weak	1
Cool	Normal	Weak	1
Cool	Normal	Strong	2
Mild	Normal	Weak	1
Mild	High	Strong	2

All records in Table 4.4 belong to the same class, hence it becomes a leaf with class label = 1 (Yes). This leaf is reached by branch *Outlook = Overcast.*

Next, computing CVGain using the Table 4.5 (node along the branch *Outlook = Rain*), the attribute *Wind* has highest CVGain when compared to *Temperature* and *Humidity*. So, two branches are grown from this node, leading to two leaves. The leaf with class label = 1 (Yes) is formed along the branch *Wind = Weak* and the other leaf with class label = 2 (No) is along the branch *Wind = Strong*. This decision tree is shown in Figure 4.1.

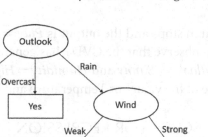

FIGURE 4.1   CVDT for data in Table 4.2.

Classification rules can be derived from this decision tree by traversing from root to leaves. The rules from the CVDT of Figure 4.1 are given below.

```
IF Outlook=Sunny AND Humidity=Normal
        THEN 'Play = Yes'
IF Outlook=Sunny AND Humidity=High
        THEN 'Play = No'
IF Outlook=Overcast
        THEN 'Play = Yes'
IF Outlook=Rain AND Wind = Weak
        THEN 'Play = Yes'
IF Outlook=Rain AND Wind = Strong
        THEN 'Play = No'
```

4.6.3 Using CVDT for Classification

Algorithm 7 will construct the decision tree during training phase. This tree will be used to predict the class label of new data objects. For a new object, traverse the tree from root node to leaf based on the values of the features. For example, if the object is *Outlook = Sunny, Temperature = Mild, Humidity = High*, and *Wind = Strong*, then we use the edge *Outlook = Sunny* to reach the left child of root node. Then, using edge *Humidity = High*, the right child of this *Humidity* node is reached. As a leaf node is reached, search stops and the output is *Play* = *No*. From this example, we can observe that the CVDT has generalized the concept of *Play*. If *Outlook* = *Sunny* and *Humidity* = *High* then *Play* = *No* irrespective of the values of Temperature and Wind.

4.7 CVDT FOR REGRESSION

CVGain can be used to build the regression tree when the decision attribute is continuous. It can approximate the real valued attribute for prediction. The decision trees that can predict the decision attribute value for continuous class attributes are normally called as regression trees. As the decision attribute is numeric, there is no need to preprocess the decision attribute. The process of splitting the data to reduce the variation in data (by using CVGain) generates nodes that have least sum of squared error from the mean of the node. When CV of a node is below a threshold α, it becomes a leaf. Changing line 7 of Algorithm 7 as **if** $CV(D) < \alpha$ will create leaf nodes. In lines 8, 11, and 19, label of the leaf has to be the mean of the data. If $(X_1, y_1), (X_2, y_2), \ldots, (X_c, y_c)$ are the objects at the leaf, then the predicted value is $\frac{1}{c} \sum_{i=1}^{c} y_i$. With these changes, Algorithm 7 can build regression tree. This algorithm is described in 8.

Predicting the continuous attribute "miles per gallon" (mpg) produces the regression tree shown in Figure 4.2 when CVDT

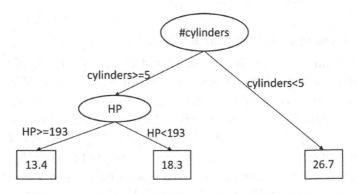

FIGURE 4.2   Regression tree of AutoMPG data set.

Algorithm 8 CVDTR: Regression Tree Using CVGain

Require: $DT, D, A = \{A_1, A_2, ...A_n\}$ ▷ D is the decision attribute, A is the set of features of Decision Table DT

Require: α ▷ Threshold for CV

1:   $N \leftarrow createNode(DT)$
2:   **if** $CV(D) < \alpha$ **then**
3:      $label(N) \leftarrow mean(D)$ ▷ CV is low, objects are homogeneous
4:      return N as leaf
5:   **end if**
6:   **if** A is empty **then**
7:      $label(N) \leftarrow mean(D)$ ▷ majority
8:      return N as leaf
9:   **end if**
10:   $A_s \leftarrow \arg\max CVGain(DT, A, D)$ ▷ A_s is the splitting attribute
11:   **if** $CVGain(A_s) < 0$ **then**   ▷ CVGain is negative, terminate with leaf
12:      $label(N) \leftarrow mean(D)$
13:      return N as leaf
14:   **end if**
15:   $A \leftarrow A - A_s$ ▷ A_s is categorical
16:   **for all** $v_j \in A_s$ **do**   ▷ partition the tuples and grow sub trees
17:      $DT_j \leftarrow$ Partition of DT with $A_s = v_j$
18:      **if** DT_j is empty **then**
19:        attach leaf node to N, labeled with most probable value in DT
20:      **else**
21:        attach tree $CVDTR(DT_j, A, D)$ to N
22:      **end if**
23:   **end for**

---

for regression is used on Auto MPG data set from UCI Machine Learning Repository [14]. This data set has 398 objects and 8 independent attributes to predict the fuel consumption of cars in *mpg*.

The threshold α for creating leaf nodes can be taken as 33%. It can even be smaller, as low as 15%. Root Mean Squared Error (RMSE) [24] is the metric for measuring the performance of regression trees. The behavior of RMSE when we change α is shown in Figure 4.3.

We can compare CVDT with *Classification and Regression Tree* (CART), which is the popular regression tree method available in all Data Analytics software packages and see if it has an improvement. This comparison is based on RMSE values. When the distributions of RMSE values of CVDT and CART are compared using Monte Carlo simulation for 100 with $\alpha = 33\%$, CVDT has mean $\mu = 38.57$ ($\sigma = 1.17$), whereas CART has mean $\mu = 45.69$ ($\sigma = 4.63$). This comparison is shown in Figure 4.4. It can be observed that CVDT has low RMSE values with less deviation.

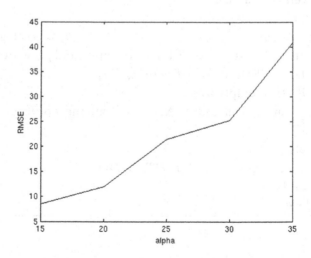

FIGURE 4.3   Variation of RMSE with alpha.

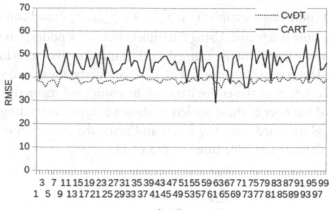

FIGURE 4.4   RMSE of CVDT and CART.

4.8 CVDT FOR BIG DATA

Distributed decision trees are needed when the data are large or when the data are distributed. Distributed decision trees available in the literature are constructed by using horizontal or vertical fragmentation of the data. CVDT can be constructed in distributed environments with these fragmentation approaches due to the algebraic characteristics of CV. CVDT can be constructed using either of the fragmentation approaches.

4.8.1 Distributed CVDT Induction with Horizontal Fragmentation

With horizontal fragmentation, subsets of data tuples are stored at different nodes. Assume that the data set $\mathcal{D}$ is available in k horizontal fragments $\mathcal{D}_1, \mathcal{D}_2, \ldots, \mathcal{D}_k$. Each node will have data for all attributes. Each node can maintain the information needed to compute CVGain as a triplet consisting of the following:

1. Number of tuples
2. Sum of decision attribute values of the tuples
3. Sum of squares of decision attribute values of the tuples

Node j computes the triplet $\left\langle n_j, \sum_{i=1}^{n_j} Y_{ji}, \sum_{i=1}^{n_j} Y_{ji}^2 \right\rangle$, and communicates only this triplet. Using the triplets from all k nodes, it is easy to compute the CV of the entire data. The number of objects in the entire data is sum of n_j for $j = 1, 2, \ldots k$. Similarly, mean and standard deviation of the entire data can be computed using the 2nd and 3rd member of these triplets as shown below. Computing CV is straightforward once the mean and standard deviation of the entire data are available (use equation 4.1).

$$n = \sum_{j=1}^{k} n_j$$

$$\mu = \frac{1}{n} \sum_{j=1}^{k} \sum_{i=1}^{n_j} Y_{ji}$$

$$\sigma = \sqrt{\frac{1}{n} \left(\sum_{j=1}^{k} \sum_{i=1}^{n_j} Y_{ji}^2 \right) - \mu^2}$$

To compute conditional CV of an attribute using equation 4.2, for each value in the domain of the attribute, the triplet has to be computed. This facilitates the CVGain computation for all the attributes and hence the CVDT induction for the distributed data.

4.8.2 Distributed CVDT Induction with Vertical Fragmentation

With vertical fragmentation, each node will have all objects with a subset of attributes, i.e., n objects with only some of the attributes. Each node will have a subset of conditional attributes and the decision attribute. The CV of the entire data is computed using the triplet of any node (as each node has all tuples for the decision attribute). Each node will compute the CV and conditional CV for the attributes available at the node and shares them with the central node. The best attribute for splitting, i.e., the attribute with highest CVGain can be computed knowing the CV and conditional CV of all attributes. Thus, CVDT can be induced.

4.9 FUZZY CVDT

In applications like Control Problems, usually the class to which an object belongs is fuzzy. Forcing crisp classification in such situations makes us lose meaningful interpretation. Generating a membership function determining the degree to which the object belongs to each of the classes is useful in such applications. Given a set of classes $\mathcal{C}$, we wish to determine a membership function $\mu(x) \to [0, 1]$ for an object x such that $\sum_{c \in \mathcal{C}} \mu(x) = 1$. Let the data set be described as a decision table DT with the features $A_1, A_2, \ldots, A_n$ and a dependent variable Y. In this context, let our decision attribute Y consists of the set of classes $\mathcal{C} = \{\mathcal{C}_1, \mathcal{C}_2, \ldots \mathcal{C}_c\}$. Let $F = \{F_1, F_2, \ldots F_c\}$ be the fuzzy membership values corresponding to these classes. Now, we can arrive at c decision tables, one for each class $\mathcal{C}_i$, using its fuzzy membership value F_i. That is, the decision tables are $DT_1, DT_2, \ldots, DT_c$, where DT_i consists of $A_1, A_2, \ldots, A_n, F_i$. Using these c decision tables, c CVDTs are built, which will contain fuzzy membership values at the leaves. The leaves of $CVDT_i$ correspond to fuzzy memberships of class C_i, i.e., F_i. This outlined procedure develops the decision system to output fuzzy membership of every class of Y. See [25] for further details.

To use this fuzzy decision system for classification, a new data object is fed to all the c decision trees. These trees generate the fuzzy membership for each class out of $\{\mathcal{C}_1, \mathcal{C}_2, \ldots \mathcal{C}_c\}$. Then, a normalized vector of these c values is generated as output by this decision system so that $\sum_{c \in \mathbb{C}} F_c = 1$.

For the Play Tennis data shown in Table 4.6, two CVDTs are built, one for predicting the fuzzy membership of the class $Play = Yes$ (see Figure 4.5) and another for predicting the fuzzy membership of the class $Play = No$ (see Figure 4.6).

For the object *(Sunny,Cool,normal,False)*, fuzzy membership from FuzyCVDT for F_{Yes} is 0.398 and fuzzy membership from Fuzzy CVDT for F_{No} is 0.602. As can be observed from Table 4.6, the original fuzzy memberships are 0.44 and 0.56 (predictions are close to ground truth). The fuzzy rules followed for this object are

TABLE 4.6   Play Tennis Data with Fuzzy Memberships

Outlook	Temperature	Humidity	Wind	Yes_FM	No_FM
Overcast	Hot	High	Weak	0.12	0.88
Overcast	Mild	High	Strong	0.1	0.9
Overcast	Hot	Normal	Weak	0.41	0.59
Overcast	Cool	Normal	Strong	0.26	0.74
Rain	Mild	High	Weak	0.28	0.72
Rain	Mild	High	Strong	0.82	0.18
Rain	Cool	Normal	Weak	0.15	0.85
Rain	Mild	Normal	Strong	0.12	0.88
Rain	Cool	Normal	Weak	0.64	0.36
Sunny	Hot	High	Strong	0.84	0.16
Sunny	Mild	High	Weak	0.83	0.17
Sunny	Hot	High	Strong	0.58	0.42
Sunny	Cool	Normal	Weak	0.44	0.56
Sunny	Mild	Normal	Strong	0.25	0.75

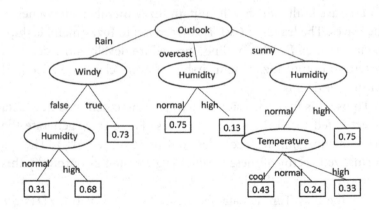

FIGURE 4.5   CVDT for fuzzy membership of *Play = Yes*.

```
IF Outlook=Sunny AND Humidity=Normal
AND Temp=Cool THEN 'Play = Yes' with
   Fyes = 0.398


IF Outlook=Sunny AND Humidity=Normal THEN
'Play = No' with Fno = 0.602
```

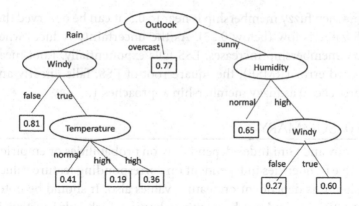

FIGURE 4.6 CVDT for fuzzy membership of *Play = No*.

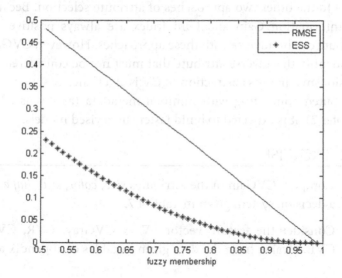

FIGURE 4.7 Fuzzy membership influence on RMSE and ESS.

The influence of fuzzy membership on the classification performance can be observed by varying the fuzzy membership from 0.51 to 0.99. Figure 4.7 shows the average *Error Sum of Squares* (ESS) behavior using the weather data set of Table 4.6 with 10 Fold Cross Validation. It is clear that the ESS decreases as fuzzy membership for a class approaches 1. Even though the uncertainty is high when fuzzy membership is near to 0.5, it can be observed that ESS value is low (below 0.25). As the uncertainty reduces when fuzzy membership increases, ESS falls exponentially. Root mean squared error (RMSE), the square root of ESS, falls linearly and approaches 0 as fuzzy membership approaches 1.

4.10 SUMMARY

Entropy and Gini Index depend only on probabilities or empirical relative frequencies independent on corresponding feature values, while CV is dependent on feature values also. It should be noted that CVGain need not be positive always, which helps further in selecting the features of dichotomy purpose more precisely compared to the other two approaches of attribute selection. Because the Information Gain and Gini Index are always positive, no attribute can be adverse with these approaches. However, CVGain can identify the adverse attribute that must not be considered for decision-making. As abstraction of CV is algebraically suitable for distributed computing with minimal metadata (as discussed in Chapter 2), it is expected to build faster supervised models.

4.11 EXERCISES

1. Compute CVGain of the attributes *size*, *color*, and *shape* for a decision system given in Table 4.7.

2. Consider the feature vector FV = CVGray, CVR, CVG, CVB, Hvalue from the data given in tables of appendix as a

TABLE 4.7   Decision System

Size	Color	Shape	Class
Medium	Blue	Brick	Yes
Small	Red	Sphere	Yes
Large	Green	Pillar	Yes
Large	Green	Sphere	Yes
Small	Red	Wedge	No
Large	Red	Wedge	No
Large	Red	Pillar	No

decision system with *Hvalue* as decision variable, build classifiers by the following methods:

a. CVDT

b. Fuzzy CVDT (use triangular fuzzification)

3. Consider the feature vector FV= with continuous values [CVGray, CVR, CVG, CVB, Hvalue] from the data given in Appendix tables as a decision system with *Hvalue* as decision variable, build classifiers by the following methods:

a. CVDT

b. Fuzzy CVDT

4. Construct a CVDT from the data given in Table 4.8. It contains generalized Employee data with number of such tuples given in *count* attribute. For example, the first row states that 30 employees of *sales* department, aged 31...35 and drawing salary in the range 46...50 thousands have *status = senior*.

5. Construct CVDT for the two-dimensional data in Figure 4.8 having circle and star classes. Infer the classification rules induced by this decision tree.

TABLE 4.8 Employee Data Set

Department	Age	Salary in Thousands	Status	Count
Sales	31...35	46...50	Senior	35
Sales	26...30	26...30	Junior	45
Sales	31...35	31...35	Junior	45
Systems	21...25	46...50	Junior	25
Systems	31...35	66...70	Senior	10
Systems	26...30	46...50	Junior	5
Systems	41...45	66...70	Senior	5
Marketing	36...40	46...50	Senior	10
Marketing	31...35	41...45	Junior	5
Secretary	46...50	36...40	Senior	5
Secretary	26...30	26...30	Junior	10

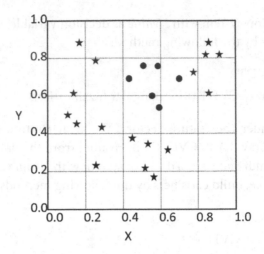

FIGURE 4.8   Two-dimensional data for classification.

6. Predict salary using GPA and experience in months for the data given in Table 4.9 by building coefficient of variation based regression tree. The salary is in thousands.

7. Construct CVDT for the data sets *Iris, Wine* and *Breast cancer Wisconsin* available at https://archive.ics.uci.edu/ml/index.php.

TABLE 4.9   Regression Data Set

GPA	Experience	Salary
2.8	48	20
3.4	24	24.5
3.2	24	23
3.8	24	25
3.2	48	20
3.4	36	22.5
4.0	20	27.5
2.6	48	19
3.2	36	24
3.8	12	28.5

Applications

SOME OF THE POTENTIAL Machine Learning applications of coefficient of variation (CV) are discussed in this chapter.

5.1 IMAGE CLUSTERING

Clustering is an unsupervised learning [2, 3] where relationships of objects in feature space are used to group related objects. Image clustering is performed on vector representation of images. CVImage features discussed in Chapter 3 can be used for performing image clustering (e.g., clustering of dog images and cat images). For example, k-means clustering [2,3] on the CV features of color images given in Appendix Tables A.1 through A.8.

Elbow method is normally used to determine the number of clusters in the given data by considering within sum of cluster variances [2]. By using the CV Features, mean, standard deviation, and CV for colors Red, Green, and Blue, a feature vector of length 9, can be formed. The result of elbow method on all color images with 9 features is shown in Figure 5.1a. It is observed that using CV of RGB colors, i.e., CV Feature vector of size 3 has a similar outcome as shown in Figure 5.1b. The best possible number of clusters is 2 from the Elbow analysis. Although the color images are from 7 databases of various types of images, they can primarily be grouped into two clusters. Further analysis

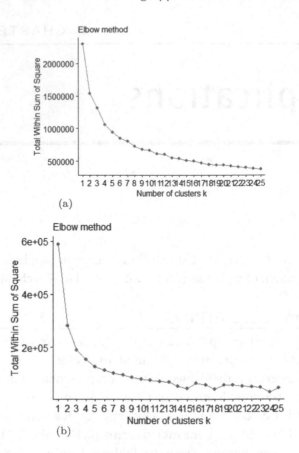

FIGURE 5.1 Elbow method on color images to determine the optimal number of clusters with CV Features. (a) Elbow method on 9 CV features. (b) Elbow method on 3 CV features.

of each group will give raise to more meaningful subgroups. The percentage of agreement between two types of clustering (k-means on 9 features and 3 features) is found to be 86%. Hence, CV Feature vector of length 3, considering only CV of RGB colors is sufficient enough for image clustering.

The result of clustering algorithm can be evaluated by intrinsic or extrinsic approaches [2, 3]. The results of extrinsic measure

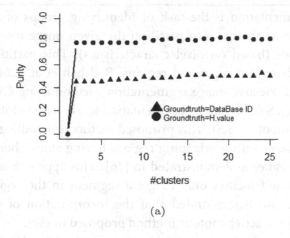

(a)

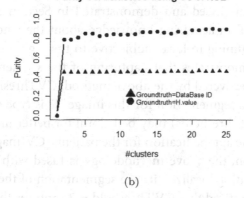

(b)

FIGURE 5.2 Purity of clusters. (a) Purity with 9 CV Features. (b) Purity with 3 CV features.

Purity [2, 3] are presented in Figure 5.2a and b, using the image database ID as ground truth and Hvalue as another ground truth. As these databases are a heterogeneous collection of images, Hvalue turns out to be better ground truth with high Purity achievement. Again, it is observed that feature vector of length 3 is sufficient from these results.

5.2 IMAGE SEGMENTATION

Image segmentation is the task of identifying groups of pixels having similar attributes to partition the given image into multiple regions (based on pixel characteristics). This partitioning is mainly based on edges and regions. Badshah et al. [26] have proposed a selective image segmentation method using CV as a fidelity term. Selective image segmentation works with points near object/region of interest. This proposed method is handling challenging images with overlapping regions having almost homogeneous intensities as demonstrated in [26]. This approach is semi supervised and focuses only to get a segment in the region of interest. It also demonstrated that the incorporation of CV is better than the active contour method proposed in [27].

An unsupervised segmentation can be arrived by developing CVImage as discussed and demonstrated in Section 3.2, which may be helpful in acquiring knowledge about the whole system instead of confining to local, subjective to the region of interest. Figure 5.3 demonstrates that capturing of the segments besides the segment recovered by the above method. CVThreshold = 3 is able to achieve segmentation for this image. This type of analysis (with different thresholds) may be helpful for better understanding to prescribe apt medication for the patients. CVImage is holistic information, the above methodology is fused with CVImage process for getting localized image segmentation of the region of interest. It is noticed that CVThreshold = 3 captures the segment of interest of the above methodology. Source for this brain tumor image is *https://healthfully.com/recognize-symptoms-brain-tumor-4506677.html.*

5.3 FEATURE SELECTION

Feature selection is used to restrict the features used for Machine Learning to relevant and nonredundant features [2], resulting in reduction of data size without degradation of accuracy. Feature selection is mainly used when number of features are more compared to number of objects, which is a challenge faced in

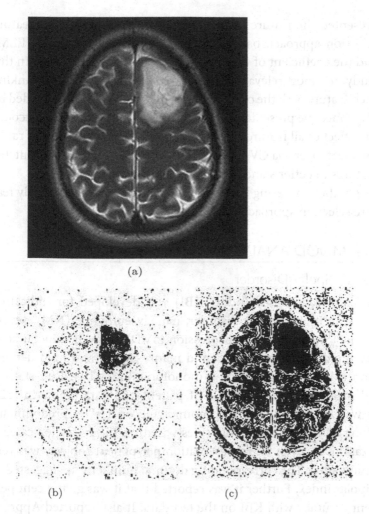

FIGURE 5.3 Image segmentation with CVImage and CVThreshold. (a) Brain tumor image; (b) result of segmentation with CVThreshold = 3; and (c) result of segmentation with CVThreshold = 9.

text mining and gene classification. Feature selection standard approaches are wrapper methods, filter methods, and embedded methods. These are based on feature ranking, the task that CV can perform efficiently. Few attempts were made to select features based on CV alone as presented in [28]. Similarly, [29] and [30] presented the feature extraction methods based on CV. Feature selection approach based on extreme learning machines (ELM) and the coefficient of variation (CV) was proposed in [31]. In this study, the most relevant features are identified by way of ranking each feature with the coefficient obtained through ELM divided by CV. Since the presented approach uses ELM, it takes into account the effect of all features and reflects onto the mechanism of ranking. Moreover, via CV values, the method selects the most suitable features to better standardize the ranking scheme.

Attribute ordering based on CV can be used in any greedy feature selection approach.

5.4 MOOD ANALYSIS

5.4.1 Bipolar Disorder

A Knowledge Based Index (KBI) was developed for indicating the severity of bipolar disorder, namely euthymic (NM), severe mania (SM), and severe depression (SVD); or cyclic bipolar state (manic depression, MDP), mild mania (MM), and mild depression (MD) by engineering a monotonic index based on statistical measures mean, SD, CV of patient's 52 data points in [32]. It was reported that a dichotomic rule on CV as predominant along with few dichotomic rules based on the mean for categorization. A seven bit representation per patient (object) was recommended. The corresponding decimal number was suggested as bipolar index. Further it was reported that it was giving cent percent accuracy with KBI on the test data. It also reported Approximate Entropy (ApEn) has very poor correlation with KBI (ApEn is not monotonic with the severity of the decease because ApEn is not designed to consider the monotonic nature of the decease severity).

5.4.2 Twitter Mood Predicts the Stock Market

An investigation of public mood correlation to Dow Jones Industrial Average (DJIA) is available in [33]. The authors of this study analyzed the tweets with two mood tracking tools, namely Opin-ionFinder that measures positive vs. negative mood and Google-Profile of Mood States (GPOMS) that measures mood in terms of 6 dimensions (Calm, Alert, Sure, Vital, Kind, and Happy). In this context, the CV can be deployed for mood indexing and mood monitoring. CV can also be used for regularization of DJIA predictive analytics.

5.5 CV FOR OPTIMIZATION

The classical balanced transportation problem solution is arrived at two phases: first phase is initial basic feasible solution and the second phase is to verify and drive toward optimal solution. The methodologies available for these phases are Greedy methodologies. "The basic transportation problem was originally developed by Hitchcock in 1941. Efficient methods for obtaining solution were developed, primarily by Dantzig in 1951 [18], followed by Charnes et al. in 1953" [11]. It is a combinatorial as well as optimization problem. The efficiency of phase 2 will depend on initial feasible solution obtained in phase 1. In soft computing paradigm, an effective initial feasible solution in phase 1 may be considered as a neat optima by avoiding phase 2. A contribution based on inverse coefficient of variation for finding an initial basic feasible solution for transportation problem has been proposed, developed, and demonstrated in [34]. The proposed Inverse Coefficient of Variation Method (ICVM) is once again a greedy approach similar to content based or data-driven methods, namely, Column Minimum Method (CMM), Least Cost Method (LCM), Row Minimum Method (RMM), Vogel's Approximation Method (VAM), and Allocation Table Method (ATM). The North West Corner Method (NWCM) is independent of content.

A window of size row/column is used as sliding window over rows. A window of size column is used as sliding window over columns. CV is computed and its inverse is used as a measure for ordering the object's rows and columns. The object with least inverse CV has been taken as a potential candidate. In the considered object, the least cost coefficient cell is taken as basic cell and maximal possible allocation has been assigned to the variable then the availability and demands are adjusted accordingly. Row/column with adjusted availability/demand as zero is removed and the reduced transportation problem is obtained as input for the next iteration. The iterations are continued until the transportation problem becomes null matrix.

As explained, CV is a measure of disorderliness and the more the CV the more the disorder. Any error in allocation of a row/column with maximal disorderliness, i.e., CV will lead to inferior solution. Hence, the CV can be treated as a penalty for corresponding row/column as in VAM. The inverse CV can be treated as a fitness, hence row/column with the least CV is best fit; hence, the above ICVM results in better initial basic feasible solution as concluded in [34] to provide near optimal solution.

5.6 HEALTH CARE

When parameters of the subject are changing drastically, CV is sensitive to disorder behavior of the disease especially in infecting stages of some contiguous diseases. CV is used for measuring the variability of quantitative bioassays like enzyme-linked immunosorbent assay (ELISA) in [35]. This work is able to assess if vaccination or infection has effect on the patient. They have come up with a mathematical relationship between CV and the frequency of disparity in bioassays when same sample is used for repeated measurements to address practical problems of clinical labs.

The medical parameters like blood pressure, sugar levels, blood density, Creatinine levels etc will fluctuate in a standard form for normal subjects while fluctuate widely for diseased subjects. CV being a measure of disorderliness, it is a potential feature for risk modelling in medical analytics.

5.7 SOCIAL NETWORK

CV of the feature values of neighborhood of a person is attributed as a node's importance or fitness in a social media network graph. Using this information, Cliques can be identified (all users with similar CV forms a Clique). Further, social network segmentation, generation of heatmaps, etc. can be performed. Each user can be represented by neighborhood CV, i.e., for every user a neighborhood ball is constructed, CV can be computed for this ball. More suitable features can be selected for social network analytics by using the mechanism available in Section 5.3.

5.8 SUMMARY

Clustering of images of considered database has been demonstrated by using CVFeatures. The potential use of CVImage in medical image segmentation, selective image segmentation, has been demonstrated.

CV-based feature ordering and feature selection, leading to a candidate order in all Data Mining and Machine Learning approaches, are proposed.

Use of CV and CVThresholds in building knowledge-based index for mood analysis is discussed. The use of inverse CV for attaining initial feasible solution for transportation problem is discussed. The applications of healthcare and social networks are discussed with some recommendations.

5.9 EXERCISES

1. Perform customer segmentation by using CV of customer expenditure patterns on the e-commerce data available at https://www.kaggle.com/fabiendaniel/customer-segmentation/data and https://archive.ics.uci.edu/ml/datasets/online+retail.

2. Perform clustering on image databases using CV flag and CV by considering the tables of appendix. What inferences can

be made with Elbow method? How is this different from performing clustering using all CV features.

3. Is Hvalue useful for hierarchical learning? Perform clustering on the objects falling into same Hvalue bin for the images data given in Appendix.

4. Compute Hvalue for the images available at http://imagedata base.cs.washington.edu/groundtruth/cherries/. What are your observations?

CV Features of Image Databases

C V FEATURES OF IMAGE DATABASES discussed in Section 3.2.2 are given in Tables A.1 through A.8. The features corresponding to standard deviation are removed from these tables due to space constraints (they can be computed when the mean and CV is known). Table A.2 is for database of grayscale images. Hence only mean and CV of the 256 images are given in this table along with image id. Hvalue if 1 for all these images as these are only gray scale images. For every feature a shorter name is used due to space constraints. The following is the mapping between short names and their meaning.

Hval: Hash Value

μ_{Gray}: Mean of Gray

CVGray: CV of Gray

CVGr: CV of Gray

CVGrF: CV Flag of Gray

μ_R: Mean of Red

CVR: CV of Red

CVRF:	CV Flag of Red
μ_G:	Mean of Green
CVG:	CV of Green
CVGF:	CV Flag of Green
μ_B:	Mean of Blue
CVB:	CV of Blue
CVBF:	CV Flag of Blue

TABLE A.1　CV Features of Greenlake

Hval	μ_{Gray}	CVGray	CVGrF	μ_R	CVR	CVRF	μ_G	CVG	CVGF	μ_B	CVB	CVBF
7	108.61	67.31	1	77.72	66.13	1	119.54	67.00	1	133.37	75.11	1
7	119.60	70.33	1	90.38	71.22	1	131.32	70.10	1	135.83	76.18	1
7	97.12	36.35	1	69.70	37.45	1	105.38	36.50	1	126.43	43.34	1
7	83.34	46.09	1	70.71	47.96	1	88.40	45.18	1	90.50	63.69	1
7	86.56	47.21	1	70.21	51.37	1	92.55	46.22	1	98.65	64.36	1
7	116.22	45.63	1	81.22	50.06	1	127.86	44.59	1	148.17	45.71	1
7	78.76	53.47	1	64.30	64.94	1	84.31	50.37	1	88.23	68.00	1
7	116.77	47.80	1	83.59	53.23	1	128.15	46.64	1	145.28	47.02	1
7	108.09	55.74	1	80.16	58.27	1	118.74	54.55	1	126.55	63.19	1
7	78.85	63.81	1	65.49	71.56	1	85.25	61.90	1	80.97	82.42	1
7	93.91	77.18	1	79.25	72.75	1	98.78	80.60	1	107.20	81.27	1
7	78.63	69.78	1	72.08	68.31	1	80.55	72.29	1	85.97	82.75	1
7	84.66	51.63	1	91.88	51.17	1	81.67	54.30	1	81.18	63.80	1
7	85.37	82.34	1	62.29	85.07	1	93.49	81.25	1	104.17	88.77	1
7	118.23	46.08	1	82.12	49.37	1	130.27	45.89	1	150.96	55.23	1
7	110.14	47.99	1	72.05	53.95	1	122.03	46.18	1	148.90	53.64	1
7	73.48	58.52	1	69.52	62.49	1	75.34	57.67	1	74.34	71.82	1
7	76.64	56.43	1	77.94	58.77	1	76.88	55.68	1	71.89	67.03	1

(Continued)

TABLE A.1 (*Continued*) CV Features of Greenlake

Hval	μ_{Gray}	CVGray	CVGrF	μ_R	CVR	CVRF	μ_G	CVG	CVGF	μ_B	CVB	CVBF
7	66.25	60.69	1	67.01	65.74	1	66.79	58.81	1	61.39	71.29	1
7	79.35	75.65	1	63.49	73.50	1	87.10	75.15	1	81.05	92.98	1
7	94.80	55.80	1	68.15	52.27	1	104.14	55.45	1	116.64	72.16	1
7	111.47	53.68	1	72.64	59.13	1	123.68	52.56	1	150.53	56.13	1
7	94.62	55.81	1	60.06	62.03	1	104.86	54.64	1	132.61	59.59	1
7	68.38	62.64	1	68.99	65.36	1	69.07	62.15	1	63.16	74.15	1
5	98.43	27.31	0	64.42	33.04	1	108.72	26.80	0	134.81	33.99	1
7	77.11	80.35	1	60.60	86.95	1	82.79	79.00	1	91.19	88.07	1
7	72.28	58.89	1	64.18	66.28	1	75.74	56.34	1	75.73	73.12	1
7	104.44	59.15	1	68.02	65.16	1	116.07	57.95	1	140.16	60.20	1
7	85.07	48.20	1	59.30	59.90	1	94.63	46.27	1	103.54	66.46	1
7	82.77	49.44	1	61.20	57.17	1	90.99	48.48	1	97.15	70.22	1
7	104.63	55.23	1	71.76	67.73	1	115.58	54.01	1	134.34	67.76	1
7	95.94	39.61	1	71.51	46.10	1	103.65	40.29	1	120.43	58.29	1
7	104.59	69.75	1	73.16	76.09	1	117.75	69.41	1	119.23	78.99	1
7	79.55	60.60	1	64.95	69.13	1	85.00	59.17	1	89.82	77.48	1

(*Continued*)

TABLE A.1 (*Continued*) CV Features of Greenlake

Hval	μ_{Gray}	CVGray	CVGrF	μ_R	CVR	CVRF	μ_G	CVG	CVGF	μ_B	CVB	CVBF
7	77.65	78.00	1	61.74	84.79	1	84.68	76.80	1	83.23	82.54	1
7	113.50	55.08	1	77.93	69.53	1	126.45	53.01	1	140.17	47.42	1
7	107.78	34.90	1	73.11	38.53	1	119.04	34.31	1	140.93	44.45	1
7	107.03	43.65	1	74.49	45.01	1	117.17	43.66	1	140.30	52.75	1
7	90.46	53.40	1	65.30	52.08	1	98.60	53.36	1	114.61	67.30	1
7	120.24	44.69	1	83.06	48.92	1	132.31	44.04	1	155.63	45.39	1
7	105.65	38.72	1	67.83	45.66	1	117.28	37.63	1	145.16	46.44	1
7	67.60	64.98	1	63.48	69.52	1	70.02	62.88	1	65.92	78.81	1
7	82.83	49.75	1	68.32	58.35	1	90.38	47.39	1	82.09	70.89	1
7	86.15	79.79	1	63.25	85.24	1	94.99	79.45	1	100.76	81.82	1
7	72.56	46.47	1	63.03	52.44	1	79.71	44.07	1	60.87	55.82	1
7	92.91	67.29	1	85.36	75.27	1	96.62	64.88	1	93.64	77.95	1
7	104.93	75.53	1	89.10	84.50	1	110.39	75.35	1	118.28	72.87	1
7	109.47	72.75	1	99.94	83.24	1	112.00	72.03	1	121.50	72.09	1

TABLE A.2 CV Features of Greenland in Gray Scale (Hvalue = 1)

S. No	meanGray	CVGr	S. No	meanGray	CVGr	S. No	meanGray	CVGr	S. No	meanGray	CVGr
1	155.05	46.25	65	126.43	70.88	129	117.83	67.31	193	145.12	60.77
2	137.75	64.90	66	136.66	54.58	130	130.99	71.26	194	122.31	59.00
3	149.11	53.02	67	145.17	56.85	131	163.44	42.60	195	113.91	73.16
4	158.05	53.81	68	163.54	47.02	132	154.78	47.38	196	148.67	51.00
5	138.66	61.12	69	151.87	55.92	133	163.51	45.81	197	156.77	49.77
6	155.19	53.50	70	155.39	56.08	134	106.11	74.97	198	129.80	55.08
7	155.19	53.50	71	141.22	68.68	135	144.35	61.74	199	143.58	51.24
8	150.90	55.93	72	140.09	66.10	136	140.30	63.78	200	160.51	52.07
9	137.47	61.80	73	126.84	81.18	137	133.64	69.14	201	169.22	47.84
10	152.88	55.88	74	133.84	50.49	138	119.74	84.07	202	159.48	54.76
11	134.35	55.42	75	142.63	59.78	139	148.46	57.09	203	168.46	48.86
12	123.13	80.49	76	144.65	59.88	140	131.95	60.16	204	122.29	56.79
13	110.95	74.37	77	147.51	58.76	141	137.53	67.82	205	163.97	48.82
14	108.00	78.18	78	130.97	65.63	142	123.45	78.72	206	80.96	98.52
15	91.44	90.68	79	128.49	75.55	143	145.39	48.07	207	94.54	90.98
16	128.59	77.55	80	131.68	65.12	144	138.96	53.50	208	114.72	70.56
17	95.42	90.38	81	106.09	79.60	145	133.38	54.32	209	158.73	53.19

(Continued)

TABLE A.2 (*Continued*) CV Features of Greenland in Gray Scale (Hvalue = 1)

S. No	meanGray	CVGr	S. No	meanGray	CVGr	S. No	meanGray	CVGr	S. No	meanGray	CVGr
18	91.55	95.67	82	118.69	75.38	146	145.26	56.83	210	158.62	44.04
19	75.02	112.91	83	100.86	91.24	147	147.38	54.67	211	155.57	53.17
20	135.83	67.41	84	140.54	51.91	148	133.23	62.65	212	148.58	60.64
21	142.73	53.08	85	131.63	50.05	149	141.50	57.81	213	145.11	67.01
22	155.53	46.89	86	142.40	55.66	150	131.84	60.87	214	148.43	52.83
23	132.87	57.97	87	139.01	56.62	151	156.27	51.33	215	145.29	50.10
24	145.34	50.97	88	134.34	57.19	152	147.34	55.80	216	144.92	50.92
25	143.51	51.29	89	137.02	56.45	153	121.01	83.77	217	145.91	45.63
26	142.22	57.02	90	140.04	54.99	154	128.85	57.61	218	143.72	55.76
27	133.18	62.66	91	143.00	58.11	155	129.48	57.76	219	141.94	64.95
28	144.84	50.56	92	136.52	62.48	156	138.64	73.25	220	139.89	45.78
29	135.53	69.25	93	131.69	53.69	157	132.84	68.28	221	154.67	56.13
30	156.74	49.14	94	163.77	45.04	158	129.55	77.11	222	75.39	121.04
31	139.23	50.91	95	138.49	61.77	159	119.05	82.45	223	82.31	112.91
32	154.61	48.86	96	155.70	52.81	160	148.23	56.93	224	91.96	91.50
33	152.54	58.26	97	132.71	61.67	161	117.78	67.45	225	112.39	73.26

(*Continued*)

TABLE A.2 (*Continued*) CV Features of Greenland in Gray Scale (Hvalue = 1)

S.No	meanGray	CVGr	S.No	meanGray	CVGr	S.No	meanGray	CVGr	S.No	meanGray	CVGr
34	151.23	63.91	98	154.84	54.17	162	130.26	64.73	226	159.99	53.75
35	155.92	51.19	99	154.80	58.38	163	143.17	56.98	227	163.03	51.80
36	125.72	75.89	100	164.86	45.30	164	140.66	58.44	228	168.47	52.67
37	103.06	86.09	101	151.14	52.28	165	143.17	55.98	229	163.59	49.65
38	136.24	48.60	102	167.37	38.70	166	142.82	56.41	230	151.79	53.63
39	135.69	54.66	103	149.09	58.32	167	121.07	56.24	231	143.97	66.06
40	139.21	54.62	104	146.81	63.06	168	131.82	65.03	232	144.95	63.66
41	133.24	63.36	105	128.10	69.42	169	139.31	53.24	233	136.15	61.93
42	128.28	65.80	106	160.24	55.91	170	144.54	54.56	234	137.90	58.56
43	133.45	59.80	107	134.85	60.14	171	137.54	56.70	235	161.59	44.94
44	144.96	44.33	108	155.58	47.88	172	144.01	54.56	236	138.59	56.86
45	131.86	52.41	109	144.03	56.71	173	135.56	61.31	237	135.95	58.91
46	139.81	46.53	110	121.13	77.73	174	153.04	49.90	238	129.62	60.74
47	140.14	44.71	111	148.74	56.44	175	150.10	54.09	239	136.20	60.21
48	129.51	60.84	112	141.89	58.02	176	135.12	61.84	240	141.14	61.65
49	128.60	60.27	113	141.92	54.79	177	135.86	60.65	241	138.61	65.61

(*Continued*)

TABLE A.2 (*Continued*) CV Features of Greenland in Gray Scale (Hvalue = 1)

S. No	meanGray	CVGr	S. No	meanGray	CVGr	S. No	meanGray	CVGr	S. No	meanGray	CVGr
50	109.36	73.47	114	141.89	58.48	178	127.77	61.27	242	119.80	62.45
51	120.99	55.19	115	141.38	42.13	179	119.59	63.16	243	126.70	60.98
52	125.92	53.00	116	142.16	56.53	180	117.40	75.10	244	129.45	46.57
53	124.28	53.94	117	137.39	65.19	181	144.82	62.16	245	125.43	60.33
54	142.25	42.51	118	132.71	66.77	182	141.58	46.19	246	143.04	53.53
55	137.29	43.08	119	141.54	53.88	183	139.54	71.03	247	147.28	44.31
56	129.83	61.85	120	139.26	54.34	184	136.13	74.41	248	111.40	68.95
57	116.61	65.68	121	128.73	69.14	185	113.56	67.81	249	136.38	57.76
58	143.81	57.20	122	143.30	57.01	186	165.06	50.09	250	113.32	85.01
59	142.14	59.68	123	146.34	54.95	187	132.27	65.02	251	129.66	55.29
60	107.08	79.23	124	101.89	71.18	188	158.92	50.91	252	130.47	55.97
61	129.41	71.80	125	118.12	75.13	189	170.44	45.63	253	117.29	69.50
62	139.63	47.56	126	117.04	59.80	190	138.67	57.05	254	131.81	57.62
63	139.61	65.71	127	139.97	58.93	191	155.18	59.29	255	137.73	51.36
64	143.30	61.20	128	118.65	67.22	192	155.25	57.58	256	138.12	49.32

TABLE A.3 CV Features of TrainDB

Hval	μ_{Gray}	CVGray	CVGrF	μ_R	CVR	CVRF	μ_G	CVG	CVGF	μ_B	CVB	CVBF
5	95.06	37.52	1	84.36	57.18	1	101.65	32.87	0	89.21	38.43	1
5	101.99	31.36	0	95.90	35.88	1	107.94	30.75	0	87.59	34.90	1
7	130.64	59.59	1	124.57	62.49	1	134.90	57.41	1	124.58	66.21	1
7	95.36	40.78	1	88.73	51.60	1	98.35	37.67	1	97.36	35.85	1
0	100.71	26.05	0	99.60	31.69	0	103.33	24.03	0	90.27	24.91	0
0	127.26	15.70	0	148.65	16.72	0	121.91	15.45	0	98.62	15.17	0
0	104.75	20.03	0	113.55	24.70	0	104.31	18.50	0	84.02	17.25	0
7	105.78	45.73	1	105.36	47.09	1	106.50	44.79	1	103.10	47.89	1
7	68.95	64.38	1	65.41	69.28	1	70.25	62.68	1	71.12	63.73	1
7	118.48	38.13	1	124.23	41.85	1	118.12	36.73	1	105.24	40.43	1
7	72.87	57.34	1	72.96	77.42	1	76.90	50.07	1	51.83	54.77	1
7	150.94	35.43	1	164.88	34.38	1	144.12	36.55	1	149.43	37.97	1
7	145.94	35.94	1	136.42	35.98	1	152.56	39.34	1	136.72	47.90	1
7	84.06	54.93	1	86.45	68.07	1	86.62	48.75	1	64.55	61.30	1
6	146.15	32.42	0	144.54	34.86	1	157.72	33.53	1	90.76	32.54	0
0	167.54	24.29	0	174.52	23.09	0	174.95	28.95	0	111.06	26.92	0
7	138.95	46.38	1	128.67	50.42	1	144.92	45.21	1	135.12	43.43	1
7	91.98	48.44	1	93.91	54.68	1	94.25	47.67	1	75.34	55.94	1
7	153.55	47.16	1	152.29	56.46	1	155.03	47.11	1	149.15	35.51	1

(Continued)

TABLE A.3 (*Continued*) CV Features of TrainDB

Hval	μ_{Gray}	CVGray	CVGrF	μ_R	CVR	CVRF	μ_G	CVG	CVGF	μ_B	CVB	CVBF
7	109.45	50.05	1	113.85	60.66	1	118.71	47.18	1	50.25	80.85	1
7	73.38	82.69	1	65.18	110.49	1	80.29	72.44	1	59.31	84.65	1
7	62.72	110.31	1	62.94	114.42	1	63.74	108.59	1	56.92	113.18	1
7	59.02	71.53	1	96.21	65.58	1	47.25	104.50	1	22.18	126.04	1
7	106.52	43.33	1	88.30	38.25	1	110.17	45.32	1	135.63	48.29	1
0	116.25	25.87	0	118.71	29.06	0	118.41	24.44	0	98.70	28.47	0
0	91.66	20.90	0	103.25	22.23	0	92.87	19.44	0	55.00	30.71	0
1	101.16	29.73	0	108.81	32.76	0	103.60	27.93	0	68.56	35.42	1
7	140.44	51.13	1	123.87	63.77	1	146.91	47.31	1	150.36	44.50	1
0	109.83	17.62	0	65.33	20.14	0	124.22	17.39	0	152.75	17.69	0
7	116.09	36.83	1	112.21	36.85	1	118.07	36.74	1	116.06	37.94	1
7	152.04	40.05	1	145.51	36.93	1	155.14	41.38	1	153.04	46.19	1
7	75.20	45.07	1	65.13	55.37	1	85.02	44.29	1	51.03	66.01	1
7	90.14	61.17	1	109.95	50.64	1	86.62	67.92	1	56.38	94.43	1
7	105.30	70.44	1	92.83	80.70	1	108.91	68.79	1	119.43	64.66	1
7	123.92	55.08	1	142.75	54.49	1	121.73	55.69	1	85.80	73.13	1
7	106.71	53.62	1	129.79	49.31	1	103.84	54.29	1	60.80	80.25	1
7	73.10	66.24	1	63.37	72.54	1	77.87	64.71	1	74.41	64.53	1

(*Continued*)

TABLE A.3 (*Continued*) CV Features of TrainDB

Hval	μ_{Gray}	CVGray	CVGrF	μ_R	CVR	CVRF	μ_G	CVG	CVGF	μ_B	CVB	CVBF
7	113.19	63.71	1	112.74	68.08	1	115.23	61.91	1	103.85	70.25	1
7	113.92	62.24	1	117.92	65.90	1	118.79	60.70	1	78.36	71.62	1
0	145.48	29.14	0	126.97	31.40	0	157.92	27.93	0	130.02	31.47	0
7	134.82	34.27	1	136.91	36.33	1	136.30	34.14	1	121.54	36.69	1
7	65.69	93.83	1	67.22	113.56	1	67.74	89.25	1	51.27	84.67	1
7	80.42	68.47	1	92.62	70.42	1	81.30	67.91	1	44.00	101.03	1
7	122.60	56.82	1	126.60	57.90	1	126.51	56.10	1	91.85	60.93	1
7	45.79	90.98	1	17.96	238.02	1	52.05	89.54	1	86.77	43.71	1
7	127.84	52.99	1	131.88	54.39	1	125.51	52.20	1	129.26	54.30	1
7	101.59	60.64	1	103.59	65.07	1	104.76	58.79	1	80.01	64.41	1
7	80.79	58.60	1	91.60	55.30	1	81.04	60.55	1	51.17	79.34	1
7	134.96	41.66	1	124.44	44.38	1	145.36	40.77	1	109.09	44.18	1
7	92.20	41.77	1	86.62	44.43	1	101.45	41.41	1	59.17	41.75	1
0	156.77	16.14	0	175.11	16.28	0	153.95	16.98	0	123.04	27.47	0
1	121.25	27.52	0	120.26	27.92	0	128.49	27.75	0	86.71	35.10	1
5	102.81	37.84	1	101.80	43.03	1	110.29	32.12	0	66.91	75.77	1
7	71.56	72.30	1	82.16	68.92	1	72.45	72.03	1	39.25	102.91	1
7	91.64	42.17	1	88.63	45.44	1	95.68	40.94	1	78.76	51.59	1

(*Continued*)

TABLE A.3 (*Continued*) CV Features of TrainDB

Hval	μ_{Gray}	CVGray	CVGrF	μ_R	CVR	CVRF	μ_G	CVG	CVGF	μ_B	CVB	CVBF
7	106.52	58.94	1	123.16	51.72	1	102.31	64.53	1	84.46	66.99	1
3	158.46	33.30	1	188.68	30.58	0	155.15	35.11	1	96.23	49.39	1
0	145.28	23.99	0	138.61	24.92	0	146.60	24.81	0	156.02	29.59	0
7	81.90	48.61	1	88.13	53.24	1	83.13	45.14	1	59.13	61.66	1
7	90.69	46.75	1	84.12	52.86	1	97.91	42.55	1	70.78	66.66	1
7	126.33	56.28	1	101.13	60.55	1	142.62	56.59	1	108.46	65.47	1
6	116.78	51.23	1	80.00	75.98	1	130.74	48.20	1	141.24	31.79	0
1	96.19	31.82	0	93.81	30.06	0	99.81	31.88	0	83.70	48.52	1
7	75.63	77.46	1	76.33	87.00	1	75.80	74.76	1	72.90	71.33	1
7	111.60	44.20	1	129.75	37.47	1	105.42	47.37	1	95.79	57.99	1
7	134.21	61.58	1	133.99	62.40	1	135.39	61.01	1	128.49	64.54	1
1	135.28	26.80	0	138.40	25.97	0	137.38	25.51	0	116.13	39.61	1
0	148.43	26.83	0	159.45	25.59	0	147.92	27.65	0	121.81	27.35	0
7	101.59	59.59	1	95.22	67.00	1	107.77	55.53	1	86.44	67.94	1
7	90.82	56.80	1	98.20	53.54	1	91.65	60.63	1	67.08	71.24	1
7	110.16	41.44	1	133.37	40.40	1	106.48	42.92	1	68.13	48.42	1
7	130.26	37.47	1	122.63	42.41	1	135.89	36.54	1	121.06	41.21	1
7	89.02	50.79	1	106.35	48.22	1	89.81	50.99	1	39.58	83.69	1

(*Continued*)

TABLE A.3 (*Continued*) CV Features of TrainDB

Hval	μ_{Gray}	CVGray	CVGrF	μ_R	CVR	CVRF	μ_G	CVG	CVGF	μ_B	CVB	CVBF
7	91.43	52.86	1	78.10	56.35	1	95.47	53.53	1	105.66	48.26	1
4	92.65	31.21	0	87.66	33.58	1	94.41	32.60	0	96.51	24.32	0
6	104.25	35.22	1	100.69	34.62	1	105.25	37.32	1	108.32	27.17	0
7	111.23	69.04	1	130.93	64.28	1	103.39	75.46	1	99.94	70.31	1
7	133.26	48.02	1	147.77	46.33	1	138.42	49.20	1	68.61	55.99	1
7	73.20	57.61	1	73.76	61.73	1	77.79	53.80	1	48.19	76.29	1
7	149.41	42.30	1	181.66	34.91	1	144.30	44.98	1	91.02	68.41	1
7	79.72	79.82	1	85.48	89.03	1	80.02	79.95	1	63.36	83.98	1
7	88.24	72.75	1	93.83	72.06	1	89.75	72.29	1	65.61	86.90	1
7	119.08	37.18	1	94.22	40.32	1	130.99	36.38	1	123.07	42.60	1
7	91.55	57.71	1	104.45	53.44	1	90.59	59.43	1	62.61	70.03	1
5	64.75	35.32	1	65.87	41.81	1	73.22	29.55	0	18.09	119.55	1
7	118.48	40.36	1	98.59	43.62	1	130.42	40.33	1	109.31	36.89	1
3	106.16	36.65	1	91.11	31.84	0	111.36	36.58	1	118.95	53.81	1
7	129.07	39.73	1	137.91	35.57	1	131.42	38.77	1	93.88	68.73	1
7	115.77	36.59	1	109.62	39.55	1	120.27	35.29	1	108.72	56.92	1
7	115.10	36.36	1	116.54	35.79	1	115.03	36.88	1	111.57	42.05	1
7	82.23	75.19	1	79.18	81.22	1	83.78	74.42	1	82.25	73.36	1
4	128.38	35.42	1	91.43	48.62	1	141.22	32.60	0	159.32	32.49	0

(*Continued*)

TABLE A.3 (*Continued*) CV Features of TrainDB

Hval	μ_{Gray}	CVGray	CVGrF	μ_R	CVR	CVRF	μ_G	CVG	CVGF	μ_B	CVB	CVBF
7	132.83	59.91	1	117.89	69.88	1	139.57	56.65	1	137.26	61.71	1
6	124.25	46.99	1	94.34	70.35	1	130.58	45.09	1	170.19	23.65	0
0	123.42	17.87	0	126.36	17.63	0	123.07	18.00	0	117.37	23.70	0
7	130.17	49.45	1	119.58	44.63	1	133.93	50.87	1	138.53	55.99	1
7	121.83	51.75	1	124.76	48.50	1	121.43	52.12	1	115.98	61.48	1
7	138.51	34.47	1	139.32	33.37	1	139.96	34.19	1	128.88	43.75	1
7	99.21	68.25	1	104.05	67.03	1	101.31	68.25	1	75.80	96.37	1
7	132.39	50.78	1	139.26	50.79	1	136.22	50.55	1	94.53	64.83	1
7	93.13	56.45	1	78.80	61.32	1	99.65	50.53	1	97.24	87.96	1
1	116.54	28.81	0	132.61	27.47	0	115.93	29.31	0	77.44	34.81	1
5	66.52	36.35	1	48.62	52.54	1	67.18	30.71	0	110.13	37.68	1
7	162.12	41.57	1	175.37	36.48	1	159.34	45.59	1	141.89	49.10	1
7	86.31	64.87	1	114.74	57.13	1	71.25	83.66	1	89.18	66.25	1
7	125.98	36.98	1	123.70	41.07	1	126.18	37.44	1	130.98	45.44	1
7	156.72	37.51	1	136.35	38.29	1	161.00	38.60	1	188.10	33.91	1
7	123.63	52.81	1	143.96	52.19	1	124.92	54.29	1	63.57	69.59	1
1	188.70	32.04	0	187.43	30.50	0	190.81	32.39	0	181.02	38.09	1
7	140.53	38.87	1	153.73	41.79	1	137.84	40.62	1	119.78	45.66	1

(*Continued*)

TABLE A.3 (*Continued*) CV Features of TrainDB

Hval	μ_{Gray}	CVGray	CVGrF	μ_R	CVR	CVRF	μ_G	CVG	CVGF	μ_B	CVB	CVBF
7	74.27	42.75	1	109.16	39.19	1	64.81	46.48	1	31.35	84.84	1
7	114.07	39.12	1	101.93	37.72	1	121.54	39.12	1	107.49	53.66	1
1	150.60	29.14	0	116.28	27.98	0	160.74	30.07	0	188.27	35.67	1
7	135.98	42.13	1	117.95	49.08	1	148.11	40.99	1	120.83	39.48	1
7	106.82	56.56	1	136.55	53.02	1	103.18	61.91	1	47.70	108.38	1
1	175.35	29.87	0	165.28	27.43	0	178.91	29.11	0	183.30	42.35	1
7	129.79	34.69	1	137.63	40.67	1	128.09	33.29	1	118.00	33.72	1
7	132.79	33.87	1	147.82	35.16	1	131.76	33.75	1	98.63	36.95	1
7	102.61	55.46	1	99.93	59.20	1	104.21	54.91	1	101.41	51.61	1
7	96.78	84.06	1	90.74	92.99	1	98.34	82.01	1	104.63	74.49	1
5	118.07	32.60	0	113.48	34.27	1	122.99	31.73	0	104.61	39.28	1
7	100.74	49.45	1	97.42	57.69	1	105.38	46.59	1	85.24	60.72	1
7	155.69	36.64	1	195.40	34.21	1	145.07	38.11	1	106.06	41.31	1
7	140.01	40.96	1	163.54	41.41	1	133.11	39.75	1	113.88	47.63	1
7	113.73	40.01	1	104.10	44.05	1	119.08	37.43	1	111.44	48.18	1
7	96.37	43.51	1	87.58	54.06	1	101.13	39.96	1	94.74	46.82	1
7	99.37	37.16	1	91.98	44.09	1	111.19	33.43	1	57.87	52.45	1
0	134.89	12.45	0	135.02	12.68	0	146.16	12.43	0	76.62	20.93	0

(*Continued*)

TABLE A.3 (*Continued*) CV Features of TrainDB

Hval	μ_{Gray}	CVGray	CVGrF	μ_R	CVR	CVRF	μ_G	CVG	CVGF	μ_B	CVB	CVBF
7	61.53	65.75	1	64.95	77.01	1	62.40	65.75	1	48.30	66.43	1
7	155.79	39.21	1	136.82	39.22	1	169.35	38.12	1	135.76	49.65	1
7	127.45	38.46	1	138.28	33.49	1	123.79	40.62	1	117.88	44.21	1
7	104.98	36.59	1	132.91	33.75	1	99.59	37.69	1	59.35	48.05	1
4	123.97	28.74	0	120.50	40.05	1	131.03	27.40	0	96.77	27.98	0
7	66.94	71.53	1	56.06	85.14	1	72.40	66.49	1	67.39	81.99	1
7	120.53	60.94	1	149.46	54.28	1	109.29	65.53	1	102.37	66.33	1
5	112.44	32.36	0	117.27	33.14	1	120.67	31.12	0	57.39	54.06	1
0	172.64	21.83	0	162.82	22.69	0	177.32	22.16	0	174.19	21.41	0
7	90.63	58.66	1	85.30	59.55	1	96.12	55.46	1	76.37	82.59	1
7	140.22	35.88	1	116.71	38.42	1	157.72	34.26	1	111.86	42.67	1
4	125.08	33.74	1	112.01	37.69	1	129.26	32.40	0	137.44	32.06	0
7	61.99	96.40	1	71.24	109.60	1	61.75	91.38	1	38.42	101.53	1
7	119.19	72.90	1	115.71	75.68	1	126.13	71.21	1	92.42	80.26	1
1	119.71	32.69	0	137.72	32.65	0	114.24	32.57	0	100.65	35.95	1
7	119.45	36.89	1	121.82	36.92	1	121.22	35.85	1	104.22	45.77	1
7	81.27	53.94	1	76.12	57.50	1	88.93	50.92	1	55.29	83.53	1
7	124.60	51.95	1	115.23	60.43	1	136.02	47.96	1	90.40	67.70	1
5	68.20	31.70	0	52.07	52.14	1	85.54	26.51	0	21.31	72.90	1

(*Continued*)

TABLE A.3 (*Continued*) CV Features of TrainDB

Hval	μ_{Gray}	CVGray	CVGrF	μ_R	CVR	CVRF	μ_G	CVG	CVGF	μ_B	CVB	CVBF
1	108.17	29.06	0	93.06	32.71	0	128.96	28.51	0	40.86	48.88	1
1	114.34	30.05	0	136.39	28.94	0	112.20	31.06	0	67.41	56.69	1
7	64.39	75.69	1	81.72	94.32	1	59.90	72.69	1	42.23	77.47	1
7	95.30	54.77	1	95.40	60.47	1	97.05	51.80	1	85.97	62.54	1
7	113.54	61.16	1	104.65	68.06	1	120.61	59.07	1	100.37	61.78	1
7	108.38	62.53	1	114.39	64.53	1	110.32	61.75	1	82.67	75.58	1
7	81.80	63.42	1	70.00	76.74	1	90.93	57.52	1	66.00	73.69	1
7	95.28	61.00	1	68.74	93.88	1	104.95	56.58	1	115.21	59.23	1
0	119.81	23.96	0	143.74	31.57	0	111.96	23.60	0	97.34	25.84	0
7	120.63	53.24	1	136.40	46.16	1	118.99	56.74	1	87.69	66.75	1
3	108.73	34.25	1	116.42	31.19	0	109.32	33.31	1	85.49	61.25	1
1	126.55	26.24	0	120.44	32.70	0	133.54	22.62	0	106.61	45.08	1
7	100.13	62.77	1	127.73	60.40	1	92.90	64.64	1	64.92	75.55	1
5	149.24	31.89	0	153.52	34.72	1	157.58	32.43	0	95.04	52.32	1
7	130.23	38.67	1	136.11	35.28	1	131.63	37.95	1	107.53	58.68	1
7	133.55	36.46	1	145.37	34.63	1	132.33	37.12	1	108.75	42.51	1
7	106.56	52.03	1	117.28	58.31	1	107.48	49.88	1	73.60	45.04	1
7	93.40	62.19	1	103.64	62.10	1	92.67	61.40	1	70.37	68.72	1

(*Continued*)

TABLE A.3 (*Continued*) CV Features of TrainDB

Hval	μ_{Gray}	CVGray	CVGrF	μ_R	CVR	CVRF	μ_G	CVG	CVGF	μ_B	CVB	CVBF
7	41.08	103.60	1	47.31	96.95	1	40.57	103.11	1	27.50	138.35	1
7	33.40	115.16	1	32.22	124.04	1	35.33	109.63	1	26.56	131.84	1
7	54.20	53.27	1	53.13	61.53	1	57.06	49.06	1	42.33	63.83	1
7	51.65	55.66	1	66.18	55.39	1	48.65	56.46	1	28.93	83.34	1
7	90.29	41.03	1	53.34	60.64	1	107.17	39.20	1	100.52	40.49	1
7	129.94	51.66	1	130.64	63.88	1	134.28	47.64	1	105.87	43.29	1
7	117.38	40.95	1	89.92	53.96	1	131.14	38.53	1	118.75	46.90	1
7	89.16	53.42	1	105.64	51.33	1	86.94	53.03	1	57.31	68.30	1
7	121.33	57.94	1	126.74	61.39	1	124.25	56.15	1	91.92	58.38	1
7	78.04	64.88	1	96.13	66.33	1	76.12	63.18	1	40.53	83.69	1
7	112.41	47.12	1	121.02	48.66	1	114.37	46.26	1	79.70	52.94	1
7	54.55	105.46	1	56.20	108.82	1	56.54	102.99	1	40.78	116.33	1
0	122.17	14.69	0	99.14	18.07	0	124.45	14.68	0	171.18	12.51	0
0	177.24	25.01	0	160.40	29.58	0	181.54	24.81	0	199.39	18.45	0
7	121.05	64.63	1	126.17	66.32	1	120.13	64.18	1	112.28	65.74	1
7	85.00	65.31	1	76.04	70.52	1	91.06	60.90	1	77.23	88.25	1
7	110.52	68.45	1	143.53	65.38	1	110.18	73.23	1	25.69	135.46	1
7	128.18	49.96	1	154.60	44.52	1	123.98	53.25	1	80.39	61.27	1

(*Continued*)

TABLE A.3 (Continued) CV Features of TrainDB

Hval	μ_{Gray}	CVGray	CVGrF	μ_R	CVR	CVRF	μ_G	CVG	CVGF	μ_B	CVB	CVBF
7	116.70	42.88	1	128.69	42.18	1	117.97	43.93	1	78.73	46.14	1
7	71.24	82.38	1	87.46	84.51	1	70.82	83.12	1	31.03	89.77	1
7	81.28	43.74	1	82.44	48.26	1	89.28	38.73	1	36.94	93.21	1
7	119.45	40.01	1	113.24	50.71	1	125.21	37.81	1	106.04	34.22	1
7	108.14	36.50	1	86.32	42.88	1	115.25	35.74	1	128.90	44.89	1
7	108.51	35.36	1	89.93	36.95	1	115.68	34.28	1	120.36	41.36	1
7	93.96	69.41	1	95.17	62.42	1	96.96	70.62	1	75.32	90.35	1
7	128.85	49.19	1	126.51	51.72	1	129.55	49.93	1	131.40	43.77	1
0	87.69	13.67	0	78.57	15.64	0	96.13	13.58	0	68.14	16.63	0
7	111.63	39.35	1	108.84	37.23	1	112.22	39.41	1	115.19	45.76	1
7	104.08	54.12	1	105.58	55.57	1	108.61	51.87	1	76.86	68.72	1
7	135.24	37.97	1	146.63	37.60	1	135.09	38.14	1	106.12	42.55	1
7	106.99	44.32	1	105.33	44.25	1	107.83	45.01	1	107.06	44.65	1
7	78.61	42.04	1	75.86	49.89	1	84.86	38.11	1	53.43	55.73	1
1	141.20	31.79	0	131.39	30.27	0	144.01	32.52	0	152.42	34.98	0
7	82.81	56.71	1	99.10	60.36	1	75.23	61.43	1	79.00	53.91	1
7	150.88	39.63	1	144.91	37.92	1	152.36	39.50	1	159.16	48.49	1

TABLE A.4 CV Features of TestDB

Hval	μ_{Gray}	CVGray	CVGrF	μ_R	CVR	CVRF	μ_G	CVG	CVGF	μ_B	CVB	CVBF
7	95.65	64.81	1	87.92	70.37	1	102.45	62.92	1	81.02	65.90	1
7	152.58	48.21	1	153.07	50.09	1	156.03	45.82	1	133.57	64.88	1
7	113.45	48.35	1	93.74	57.33	1	120.78	45.92	1	127.35	63.69	1
7	90.26	40.17	1	78.83	44.91	1	95.59	38.22	1	92.70	44.77	1
1	182.46	31.01	0	189.68	28.59	0	181.58	31.85	0	168.01	37.16	1
7	169.06	36.35	1	155.14	37.66	1	173.69	36.08	1	181.74	35.42	1
7	80.04	50.45	1	79.11	60.84	1	83.42	47.21	1	65.07	49.68	1
7	74.00	58.53	1	66.69	75.14	1	81.05	52.16	1	56.88	63.47	1
7	64.64	74.44	1	67.26	80.05	1	64.82	72.24	1	56.89	74.51	1
7	98.77	35.74	1	99.61	42.02	1	99.89	33.55	1	90.57	34.11	1
7	104.32	69.83	1	101.09	72.92	1	105.01	69.86	1	109.32	64.34	1
7	94.56	35.77	1	118.59	43.65	1	86.33	37.30	1	73.95	42.72	1
7	119.08	36.04	1	121.93	38.70	1	117.83	34.74	1	117.99	36.52	1
7	91.90	45.04	1	74.55	59.42	1	98.09	43.50	1	105.42	44.61	1
7	91.87	63.25	1	97.41	65.92	1	92.18	62.40	1	75.61	62.69	1
7	103.49	52.79	1	98.12	52.55	1	108.86	51.06	1	90.06	64.87	1
7	71.32	72.23	1	50.40	92.51	1	78.31	69.18	1	90.50	66.44	1
7	88.50	47.18	1	75.88	55.64	1	83.97	49.16	1	144.89	35.08	1
7	111.28	54.88	1	128.04	49.33	1	105.97	65.42	1	94.72	64.62	1
7	107.36	60.39	1	56.76	93.29	1	126.24	56.39	1	142.78	58.54	1

(Continued)

TABLE A.4 (Continued) CV Features of TestDB

Hval	μ_{Gray}	CVGray	CVGrF	μ_R	CVR	CVRF	μ_G	CVG	CVGF	μ_B	CVB	CVBF
7	111.72	69.35	1	107.81	76.81	1	119.64	65.90	1	81.27	76.02	1
7	93.02	66.71	1	116.49	63.20	1	89.00	66.71	1	52.15	92.40	1
7	86.63	59.83	1	93.21	64.52	1	88.40	59.15	1	60.29	60.83	1
7	152.25	38.66	1	154.44	37.36	1	154.54	39.09	1	134.74	47.44	1
7	88.45	80.50	1	96.60	84.11	1	90.75	77.45	1	55.22	93.92	1
7	89.17	53.71	1	99.11	56.77	1	92.21	49.96	1	47.54	78.54	1
7	107.79	54.37	1	109.02	56.31	1	111.81	53.64	1	83.81	59.35	1
7	83.25	38.21	1	91.04	41.98	1	86.94	33.50	1	43.77	80.24	1
7	103.95	109.64	1	101.04	113.87	1	104.46	109.14	1	108.81	102.10	1
7	105.00	62.17	1	87.60	81.08	1	108.55	60.10	1	132.40	46.60	1
1	117.49	30.53	0	128.88	31.97	0	121.38	29.57	0	67.52	37.84	1
7	99.22	41.69	1	114.98	40.22	1	97.83	41.39	1	64.94	56.34	1
5	108.26	34.46	1	113.29	37.79	1	111.06	32.72	0	80.68	40.66	1
7	127.64	41.70	1	117.89	46.44	1	133.24	40.23	1	124.21	45.55	1
7	134.41	54.21	1	151.74	47.33	1	133.39	56.61	1	94.37	68.65	1
7	84.25	55.13	1	76.62	65.84	1	89.34	52.16	1	78.06	56.94	1
0	129.57	12.97	0	138.64	12.57	0	129.44	12.88	0	105.43	15.43	0
7	126.10	50.53	1	129.81	48.22	1	126.84	50.12	1	112.62	62.30	1
7	88.06	55.16	1	79.01	71.35	1	97.75	51.67	1	61.92	52.58	1

(Continued)

TABLE A.4 (Continued) CV Features of TestDB

Hval	μ_{Gray}	CVGray	CVGrF	μ_R	CVR	CVRF	μ_G	CVG	CVGF	μ_B	CVB	CVBF
7	105.64	39.03	1	97.26	39.44	1	113.56	38.51	1	86.90	59.02	1
7	123.55	43.48	1	120.89	43.34	1	125.03	44.11	1	122.73	43.41	1
7	85.80	70.00	1	99.06	67.01	1	84.25	72.11	1	58.93	99.77	1
7	106.59	48.01	1	100.42	52.19	1	111.37	46.56	1	98.13	47.72	1
7	88.39	61.94	1	99.47	63.43	1	87.04	61.97	1	66.24	71.93	1
7	126.88	49.06	1	114.29	56.73	1	128.95	48.09	1	149.15	39.17	1
0	111.93	27.28	0	105.56	30.34	0	114.48	26.57	0	115.48	28.60	0
7	119.61	49.13	1	137.11	41.96	1	115.70	52.51	1	93.80	64.59	1
7	110.01	33.32	1	123.26	33.98	1	106.82	33.13	1	91.70	38.69	1
7	123.57	63.67	1	129.18	62.29	1	121.93	64.84	1	117.30	64.59	1
7	130.34	56.37	1	101.58	78.78	1	143.57	49.08	1	137.68	56.75	1
7	158.63	36.54	1	144.12	43.68	1	165.56	34.25	1	160.97	39.29	1
5	112.57	36.37	1	114.46	37.39	1	118.53	34.24	1	76.95	53.16	1
0	175.66	33.26	0	170.23	33.06	0	180.14	31.29	0	166.76	49.04	0
7	160.55	19.86	1	169.23	26.50	1	164.92	18.77	1	115.29	22.50	1
7	135.24	38.05	1	158.74	36.49	1	130.84	38.35	1	96.22	48.10	1
7	86.06	58.58	1	87.26	61.35	1	87.63	60.06	1	74.79	59.54	1
7	145.54	33.19	1	143.85	36.10	1	157.56	33.68	1	88.09	47.86	1
7	85.43	43.64	1	88.28	67.82	1	84.20	42.40	1	84.35	62.36	1
1	130.80	21.74	0	126.85	21.56	0	133.60	21.98	0	126.85	38.07	1

(Continued)

TABLE A.4 (*Continued*) CV Features of TestDB

Hval	μ_{Gray}	CVGray	CVGrF	μ_R	CVR	CVRF	μ_G	CVG	CVGF	μ_B	CVB	CVBF
7	98.25	42.63	1	101.45	40.04	1	98.17	43.55	1	90.13	56.21	1
0	103.99	21.04	0	109.04	32.68	0	106.53	19.11	0	77.74	26.78	0
7	89.34	47.38	1	63.24	70.78	1	100.11	41.65	1	102.67	48.14	1
7	56.38	113.46	1	67.50	108.36	1	51.73	117.90	1	50.49	121.71	1
7	95.45	52.00	1	97.17	56.70	1	100.43	50.02	1	65.26	55.47	1
7	107.65	47.41	1	117.87	50.02	1	105.57	46.47	1	91.41	45.99	1
7	73.41	87.54	1	62.05	95.15	1	79.39	86.27	1	72.48	85.02	1
0	118.26	21.73	0	113.97	24.37	0	118.28	20.81	0	129.31	21.13	0
7	115.85	51.41	1	119.64	51.99	1	118.97	49.82	1	89.72	61.15	1
7	131.96	69.09	1	117.47	78.89	1	138.92	65.34	1	133.97	68.40	1
7	79.74	51.19	1	79.94	54.56	1	84.64	49.93	1	54.31	60.25	1
5	90.23	32.26	0	90.38	34.97	1	95.47	32.16	0	63.08	35.97	1
5	127.21	34.09	1	124.82	37.66	1	133.87	32.05	0	99.19	46.80	1
7	87.14	39.91	1	69.74	47.74	1	94.98	38.32	1	92.55	37.06	1
7	145.66	44.12	1	149.10	42.79	1	150.11	44.16	1	113.77	54.43	1
7	101.30	63.01	1	105.94	59.58	1	100.41	65.10	1	93.58	69.84	1
7	97.51	60.41	1	111.49	58.28	1	95.90	60.22	1	68.89	71.67	1
3	136.87	35.47	1	156.99	31.81	0	132.32	36.80	1	107.62	43.13	1
7	91.96	62.05	1	93.78	62.33	1	94.36	60.53	1	74.73	71.61	1
1	166.62	31.86	0	151.21	30.18	0	175.12	32.36	0	163.25	34.85	1

(*Continued*)

TABLE A.4 *(Continued)* CV Features of TestDB

Hval	μ_{Gray}	CVGray	CVGrF	μ_R	CVR	CVRF	μ_G	CVG	CVGF	μ_B	CVB	CVBF
7	68.80	42.47	1	82.71	46.97	1	68.55	38.94	1	33.41	59.94	1
7	36.00	109.63	1	34.46	145.78	1	39.57	98.30	1	22.05	119.18	1
5	101.03	30.99	0	101.04	34.10	1	103.98	30.23	0	85.72	41.40	1
7	78.23	65.07	1	90.45	61.48	1	79.93	63.38	1	37.41	114.62	1
7	56.78	96.83	1	69.51	93.91	1	54.45	99.07	1	35.57	109.39	1
7	139.40	64.12	1	138.83	68.56	1	143.83	61.76	1	118.00	66.04	1
7	85.50	53.23	1	69.43	64.80	1	100.01	49.47	1	53.06	72.45	1
7	74.15	60.34	1	86.93	63.15	1	73.36	57.75	1	44.72	75.74	1
7	61.57	66.79	1	71.08	62.26	1	59.50	68.59	1	47.21	87.16	1
5	99.50	32.80	0	98.49	36.22	1	103.63	30.36	0	80.92	45.72	1
3	81.94	37.33	1	102.54	32.96	0	74.89	39.70	1	64.22	44.55	1
7	109.15	36.22	1	110.94	38.50	1	111.12	35.61	1	94.35	40.18	1
7	115.99	61.91	1	120.18	60.20	1	116.38	62.33	1	102.97	66.28	1
0	114.95	27.20	0	119.94	29.31	0	115.43	25.88	0	99.36	32.29	0
7	144.38	42.12	1	153.80	41.59	1	141.65	43.74	1	133.71	42.43	1
7	95.83	44.79	1	93.82	66.73	1	95.94	48.79	1	100.49	44.56	1
0	154.65	28.42	0	149.77	30.10	0	157.60	27.57	0	152.21	29.39	0
0	122.83	26.91	0	121.26	31.28	0	124.33	26.29	0	119.25	23.49	0
7	138.63	38.22	1	143.60	38.47	1	139.99	38.04	1	118.61	40.36	1
7	110.79	47.58	1	109.61	50.05	1	114.07	47.04	1	96.93	46.51	1
7	136.46	56.20	1	127.36	59.85	1	139.35	55.45	1	145.54	53.82	1

TABLE A.5 CV Features of Indonesia

Hval	μ_{Gray}	CVGray	CVGrF	μ_R	CVR	CVRF	μ_G	CVG	CVGF	μ_B	CVB	CVBF
7	86.30	80.41	1	93.11	78.17	1	85.65	80.53	1	71.84	92.40	1
7	84.04	46.00	1	85.15	51.84	1	85.70	42.85	1	72.50	50.43	1
7	131.86	59.39	1	129.01	62.99	1	133.30	58.56	1	131.73	60.95	1
7	100.15	68.61	1	102.01	69.72	1	102.49	68.56	1	83.19	70.72	1
7	132.07	45.18	1	126.19	51.60	1	135.02	42.40	1	132.23	45.39	1
7	128.68	34.98	1	127.98	37.94	1	129.93	33.93	1	124.02	39.56	1
7	126.34	72.96	1	127.72	71.72	1	127.50	71.63	1	116.60	84.95	1
7	154.76	55.02	1	152.78	56.14	1	155.89	54.26	1	154.18	56.79	1
7	78.47	87.54	1	78.69	94.65	1	80.24	83.59	1	68.79	95.54	1
7	89.86	71.15	1	103.16	74.47	1	88.88	70.99	1	60.23	72.47	1
7	76.79	56.36	1	84.50	67.08	1	76.86	55.58	1	56.29	50.58	1
7	127.33	38.43	1	131.25	39.37	1	130.14	36.80	1	102.59	58.25	1
1	138.54	28.68	0	142.25	31.23	0	142.87	27.71	0	106.47	40.25	1
7	115.96	36.86	1	117.46	40.29	1	122.08	34.41	1	80.55	68.54	1
1	154.06	32.84	0	158.83	32.36	0	162.91	29.55	0	95.90	84.34	1
7	122.03	66.67	1	125.75	63.79	1	123.93	65.29	1	102.51	89.46	1
7	106.02	75.71	1	107.37	75.63	1	106.07	76.07	1	102.12	85.61	1
7	129.28	59.72	1	135.16	55.09	1	129.37	59.86	1	113.49	78.07	1
7	94.06	40.85	1	94.30	46.40	1	96.46	38.86	1	81.06	46.20	1

(Continued)

TABLE A.5 (Continued) CV Features of Indonesia

Hval	μ_{Gray}	CVGray	CVGrF	μ_R	CVR	CVRF	μ_G	CVG	CVGF	μ_B	CVB	CVBF
7	142.03	45.64	1	141.51	46.55	1	146.89	42.30	1	118.47	71.39	1
7	62.41	103.27	1	62.82	111.07	1	63.40	100.51	1	56.23	105.30	1
7	99.82	49.80	1	92.74	61.43	1	104.74	44.85	1	93.01	63.67	1
7	123.50	63.02	1	119.99	66.66	1	125.08	62.23	1	124.60	65.96	1
1	126.43	31.92	0	132.93	31.39	0	128.04	31.95	0	101.11	45.26	1
7	126.23	47.94	1	124.61	54.50	1	127.69	45.98	1	122.87	47.36	1
1	134.21	30.83	0	134.56	32.47	0	144.31	27.88	0	81.29	64.79	1
7	145.15	41.67	1	150.86	45.67	1	147.58	39.43	1	117.67	55.30	1
7	119.00	73.85	1	120.15	76.27	1	119.59	73.36	1	112.91	82.56	1
7	117.83	56.31	1	121.12	55.62	1	118.29	56.00	1	106.84	63.66	1
7	101.57	74.31	1	102.78	76.36	1	102.61	72.73	1	93.07	81.75	1
7	139.47	50.52	1	137.51	49.33	1	140.41	50.44	1	139.48	55.48	1
7	119.99	53.18	1	118.57	59.03	1	121.35	51.18	1	116.54	51.87	1
7	95.20	117.87	1	94.78	118.87	1	95.75	116.78	1	93.63	121.17	1
7	91.23	46.74	1	94.09	46.56	1	91.88	47.68	1	80.25	45.62	1
7	81.12	90.34	1	82.69	96.64	1	84.39	86.22	1	60.23	110.41	1
7	136.80	46.80	1	135.54	48.22	1	140.57	44.11	1	120.78	62.26	1

TABLE A.6 CV Features of SampleDB

Hval	μ_{Gray}	CVGray	CVGrF	μ_R	CVR	CVRF	μ_G	CVG	CVGF	μ_B	CVB	CVBF
1	145.59	20.60	0	188.26	15.05	0	135.44	22.99	0	85.80	39.79	1
7	106.52	58.94	1	123.16	51.72	1	102.31	64.53	1	84.46	66.99	1
1	96.19	31.82	0	93.81	30.06	0	99.81	31.88	0	83.70	48.52	1
1	135.28	26.80	0	138.40	25.97	0	137.38	25.51	0	116.13	39.61	1
4	92.65	31.21	0	87.66	33.58	1	94.41	32.60	0	96.51	24.32	0
6	104.25	35.22	1	100.69	34.62	1	105.25	37.32	1	108.32	27.17	0
5	64.75	35.32	1	65.87	41.81	1	73.22	29.55	0	18.09	119.55	1
1	116.54	28.81	0	132.61	27.47	0	115.93	29.31	0	77.44	34.81	1
7	105.28	66.07	1	134.59	70.34	1	96.58	86.94	1	73.30	84.35	1
7	98.25	42.63	1	101.45	40.04	1	98.17	43.55	1	90.13	56.21	1
1	119.71	32.69	0	137.72	32.65	0	114.24	32.57	0	100.65	35.95	1
1	148.88	25.81	0	192.56	27.92	0	135.32	28.34	0	104.03	36.75	1
3	106.16	36.65	1	91.11	31.84	0	111.36	36.58	1	118.95	53.81	1
5	177.80	29.40	0	194.84	35.18	1	173.63	31.06	0	154.52	40.02	1
7	59.02	71.53	1	96.21	65.58	1	47.25	104.50	1	22.18	126.04	1
7	109.45	50.05	1	113.85	60.66	1	118.71	47.18	1	50.25	80.85	1
7	156.72	37.51	1	136.35	38.29	1	161.00	38.60	1	188.10	33.91	1
0	122.17	14.69	0	99.14	18.07	0	124.45	14.68	0	171.18	12.51	0
7	72.87	57.34	1	72.96	77.42	1	76.90	50.07	1	51.83	54.77	1
3	180.23	32.22	0	190.90	27.79	0	172.67	36.15	1	191.02	33.61	1

TABLE A.7 CV Features of Testimages

Hval	μ_{Gray}	CVGray	CVGrF	μ_R	CVR	CVRF	μ_G	CVG	CVGF	μ_B	CVB	CVBF
7	86.30	80.41	1	93.11	78.17	1	85.65	80.53	1	71.84	92.40	1
7	84.04	46.00	1	85.15	51.84	1	85.70	42.85	1	72.50	50.43	1
7	131.86	59.39	1	129.01	62.99	1	133.30	58.56	1	131.73	60.95	1
7	100.15	68.61	1	102.01	69.72	1	102.49	68.56	1	83.19	70.72	1
7	132.07	45.18	1	126.19	51.60	1	135.02	42.40	1	132.23	45.39	1
7	128.68	34.98	1	127.98	37.94	1	129.93	33.93	1	124.02	39.56	1
7	126.34	72.96	1	127.72	71.72	1	127.50	71.63	1	116.60	84.95	1
7	154.76	55.02	1	152.78	56.14	1	155.89	54.26	1	154.18	56.79	1
7	78.47	87.54	1	78.69	94.65	1	80.24	83.59	1	68.79	95.54	1
7	89.86	71.15	1	103.16	74.47	1	88.88	70.99	1	60.23	72.47	1
7	76.79	56.36	1	84.50	67.08	1	76.86	55.58	1	56.29	50.58	1
7	127.33	38.43	1	131.25	39.37	1	130.14	36.80	1	102.59	58.25	1
7	84.66	51.63	1	91.88	51.17	1	81.67	54.30	1	81.18	63.80	1
7	85.37	82.34	1	62.29	85.07	1	93.49	81.25	1	104.17	88.77	1
7	118.23	46.08	1	82.12	49.37	1	130.27	45.89	1	150.96	55.23	1
7	110.14	47.99	1	72.05	53.95	1	122.03	46.18	1	148.90	53.64	1
7	73.48	58.52	1	69.52	62.49	1	75.34	57.67	1	74.34	71.82	1
7	76.64	56.43	1	77.94	58.77	1	76.88	55.68	1	71.89	67.03	1

(Continued)

TABLE A.7 (*Continued*) CV Features of Testimages

Hval	μ_{Gray}	CVGray	CVGrF	μ_R	CVR	CVRF	μ_G	CVG	CVGF	μ_B	CVB	CVBF
7	66.25	60.69	1	67.01	65.74	1	66.79	58.81	1	61.39	71.29	1
7	79.35	75.65	1	63.49	73.50	1	87.10	75.15	1	81.05	92.98	1
7	94.80	55.80	1	68.15	52.27	1	104.14	55.45	1	116.64	72.16	1
7	111.47	53.68	1	72.64	59.13	1	123.68	52.56	1	150.53	56.13	1
7	94.62	55.81	1	60.06	62.03	1	104.86	54.64	1	132.61	59.59	1
7	68.38	62.64	1	68.99	65.36	1	69.07	62.15	1	63.16	74.15	1
5	98.43	27.31	0	64.42	33.04	1	108.72	26.80	0	134.81	33.99	1
7	77.11	80.35	1	60.60	86.95	1	82.79	79.00	1	91.19	88.07	1
7	72.28	58.89	1	64.18	66.28	1	75.74	56.34	1	75.73	73.12	1
7	104.44	59.15	1	68.02	65.16	1	116.07	57.95	1	140.16	60.20	1
7	85.07	48.20	1	59.30	59.90	1	94.63	46.27	1	103.54	66.46	1
7	82.77	49.44	1	61.20	57.17	1	90.99	48.48	1	97.15	70.22	1
7	104.63	55.23	1	71.76	67.73	1	115.58	54.01	1	134.34	67.76	1
7	95.94	39.61	1	71.51	46.10	1	103.65	40.29	1	120.43	58.29	1
7	104.59	69.75	1	73.16	76.09	1	117.75	69.41	1	119.23	78.99	1

TABLE A.8 CV Features of Images_Texture

Hval	μ_{Gray}	CVGray	CVGrF	μ_R	CVR	CVRF	μ_G	CVG	CVGF	μ_B	CVB	CVBF
5	126.15	30.06	0	105.15	36.68	1	152.18	25.64	0	47.32	72.88	1
0	123.00	0.00	0	35.00	0.00	0	177.00	0.00	0	77.00	0.00	0
7	122.79	50.50	1	113.50	58.67	1	134.04	46.29	1	89.34	66.94	1
7	110.29	76.27	1	131.75	61.23	1	104.30	82.60	1	84.84	100.27	1

References

1. E. Rich and K. Knight. *Artificial Intelligence*. 2nd edition, McGraw-Hill Higher Education, New York, 1990.
2. J. Han, M. Kamber, and J. Pei. *Data Mining: Concepts and Techniques*. 3rd edition, Morgan Kaufmann Publishers Inc., San Francisco, CA, 2011.
3. P.-N. Tan, M. Steinbach, and V. Kumar. *Introduction to Data Mining, (First Edition)*. Addison-Wesley Longman Publishing Co., Inc., Boston, MA, 2005.
4. T.H. Wonnacott and R.J. Wonnacott. *Introductory Statistics for Business and Economics with Student Workbook 4e*. Wiley, Hoboken, NJ, 1990.
5. G.W. Snecdecor and W.G. Cochran. *Statistical Methods*. Wiley, New Jersey, 1991.
6. J. Forkman. Estimator and tests for common coefficients of variation in normal distributions. *Communications in Statistics – Theory and Methods*, 38(2):233–251, 2009.
7. I.F. Blake. *An Introduction to Applied Probability*. R.E. Krieger Publishing Company, Malabar, FL, 1987.
8. P. Flach. *Machine Learning: The Art and Science of Algorithms That Make Sense of Data*. Cambridge University Press, New York, 2012.
9. Z. Pawlak. Rough sets. *International Journal of Computer & Information Sciences*, 11(5):341–356, 1982.
10. Image databases. http://imagedatabase.cs.washington.edu.
11. Lena standard test image. http://www.lenna.org/editor.html, 1973.
12. H.P. Irvin. *A Report on the Statistical Properties of the Coefficient of Variation and Some Applications*, volume 6841. All Graduate Thesis and Dissertations, 1970.
13. D.C. Montgomery and G.C. Runger. *Applied Statistics and Probability for Engineers*, 5th edition, Wiley, Hoboken, NJ, 2010.
14. D. Dua and C. Graff. UCI machine learning repository. http://archive.ics.uci.edu/ml, 2017.
15. A. Rajaraman and J. David Ullman. *Mining of Massive Datasets*. Cambridge University Press, New York, 2011.

16. T. White. *Hadoop: The Definitive Guide.* 1st edition, O'Reilly Media, Inc., Sebastopol, CA, 2009.
17. R.C. Gonzalez and R.E. Woods. *Digital Image Processing.* Prentice Hall, Upper Saddle River, NJ, 2008.
18. Greenlake image database. http://imagedatabase.cs.washington.edu/groundtruth/greenlake, 1999. Accessed July 21, 2019.
19. Greenland image database. http://imagedatabase.cs.washington.edu/groundtruth/greenland/, 2002.
20. Indonesia image database. http://imagedatabase.cs.washington.edu/groundtruth/indonesia/, 2002.
21. M. Hall, E. Frank, G. Holmes, B. Pfahringer, P. Reutemann, and I.H. Witten. The WEKA data mining software: An update. *SIGKDD Explorations,* 11(1):10–18, 2009.
22. T.M. Mitchell. *Machine Learning.* 1st edition, McGraw-Hill, Inc., New York, 1997.
23. K. Hima Bindu, K. Swarupa Rani, and C. Raghavendra Rao. Coefficient of variation based decision tree (cvdt). *International Journal of Innovative Technology and Creative Engineering,* 1(6):1–6, 2011.
24. S. Holmes. Rms error. http://statweb.stanford.edu/_susan/courses/s60/split/node60.html, 2000.
25. K. Hima Bindu and C. Raghavendra Rao. Coefficient of variation based decision tree for fuzzy classification. In *Advances in Intelligent Systems and Computing,* pp. 139–149. Springer International Publishing, 2015.
26. N. Badshah, K. Chen, H. Ali, and G. Murtaza. Coefficient of variation based image selective segmentation model using active contours. *East Asian Journal on Applied Mathematics,* 2(2):150–169, 2012.
27. C. Le Guyader and C. Gout. Geodesic active contour under geometrical conditions: theory and 3d applications. *Numerical Algorithms,* 48(1–3):105–133, 2008.
28. D. Kumar. Class specific feature selection for identity validation using dynamic signatures. *Journal of Biometrics & Biostatistics,* 4(2):1–5, 2013.
29. S. Nakariyakul and D. Casasent. Hyperspectral feature selection and fusion for detection of chicken skin tumors. In B.S. Bennedsen, Y.-R. Chen, G.E. Meyer, A.G. Senecal, and S.-I. Tu, editors, *Monitoring Food Safety, Agriculture, and Plant Health.* SPIE, 2004.
30. Md. Tarek Habib and M. Rokonuzzaman. Distinguishing feature selection for fabric defect classification using neural network. *Journal of Multimedia,* 6(5):416–421, 2011.

31. Ö. Faruk Ertugrul and M. Emin Tagluk. A fast feature selection approach based on extreme learning machine and coefficient of variation. *Turkish Journal of Electrical Engineering & Computer Sciences*, 25(1–3):3409–3420, 2017.
32. V. Sree Hari Rao, C. Raghvendra Rao, and V.K. Yeragani. A novel technique to evaluate fluctuations of mood: implications for evaluating course and treatment effects in bipolar/affective disorders. *Bipolar Disorders*, 8(5p1):453–466, 2006.
33. J. Bollen, H. Mao, and X. Zeng. Twitter mood predicts the stock market. *Journal of Computational Science*, 2(1):1–8, 2011.
34. O. Jude, O. Ben Ifeanyichukwu, I. Andrew Ihuoma, and E. Perewarebo Akpos. A new and efficient proposed approach to find initial basic feasible solution of a transportation problem. *American Journal of Applied Mathematics and Statistics*, 5(2):54–61, 2017.
35. G.F. Reed, F. Lynn, and B.D. Meade. Use of coefficient of variation in assessing variability of quantitative assays. *Clinical and Vaccine Immunology*, 9(6):1235–1239, 2002.

Index

Note: Page numbers in italic and bold refer to figures and tables, respectively.

A

absolute coefficient of variation (ACV), 4, 9, 11, 23, 25
Approximate Entropy (ApEn), 85
attribute ordering, 61–63, 85

B

Big Data, 36
binarization, 48, 55
bipolar disorder, 85–86

C

C4.5, 63
calibration, 23, 34–35
CART, 63, 70, *71*
categorical variables, 32–35
Coefficient of Variation (CV), 3–6, 9–24, 26, 29–30, 38, 44, 55, *83*, 85
    feature vector, 48–54, 79
    for optimization, 86–87
Coefficient of Variation based Decision Tree (CVDT), 63–76
    for Big Data, 71–72
    for classification, 63–68
    induction, 65, 72
    for regression, 68–71

conditional CV, 59–60
consistent, 23–24
CVFeatures, 10, 88, 91–121
CVGain, 61–62, 64–69, 72, 76
CVImage, 44–48, 82, *84*
CVImage illustration, *45*

D

decision tree, 63–64, 68

E

elbow method, 79, *80*
Entropy, 3, 31–32, 34, 38, 48, 63, 76
Error Sum of Squares (ESS), 75, *75*

F

feature selection, 63, 84–85
Fuzzy CVDT, 73–76

G

Gamma distribution, 8
Gini Index, 3, 31–32, 34, 38, 63, 76

H

health care, 87–88
highly consistent (HC), 23

highly inconsistent (HI), 24
horizontal fragmentation, 71–72
Hunt's algorithm, 63
Hvalue, 10, 49–51, **52**, 53–56, 76,
    81, 89

I

ID3, 63
image clustering, 79–81
image segmentation, 82–84
Information Gain, 63, 76
Inverse Coefficient of Variation
    Method (ICVM), 86

L

Lena image, 46–48, 54
linear transformation, 9

M

mapping method, 34, 39, 58
map-reduce strategies, 36–37
mixture distributions, 13, 15–16
moderately consistent (MC), 23
moderately inconsistent (MI), 24
mood analysis, 85–86, 88

N

normal distribution, 7–8, 16–22, 32
normalization, 5–8, 23–24

P

Poisson distribution, 8
pooled coefficient of variation,
    30–31, 39

pooled data, 28–31
pre-processing, 34, 38, 57–59

R

ranking, 53–54, 85
relative translation ratio, 10
Root Mean Squared Error (RMSE),
    70, *70, 71, 75*, 76

S

selective image segmentation, 82, *83*
social network, 88

T

table lookup method, 33–34
test data, 57, 85
train data, 57–58, **58**

U

UCI repository, 41
unsupervised learning, 79

V

vertical fragmentation, 72

W

weak consistent (WC), 23
weak inconsistent (WI), 23

Z

zero avoiding calibration, 34–35

Printed in the United States
by Baker & Taylor Publisher Services

Printed in the United States
by Baker & Taylor Publisher Services